Plum Island
4,000 on a Barrier Beach
Written by William Sargent

Table of Contents

Acknowledgements-4

Fact and Fiction-6

The Blizzards of 2015 -10

Garnet, Mica, Quartz, and all that other Schist.-14

The Glacier-18

The Matriarch-22

The Sand -26

"Rips and Runnels"-30

The Marsh-34

The Match -38

Captain John Smith-44

"The New England Prospect" -48

Wicked Walis -52

The Ladies of Grape Island -56

The Storm-60

Wits and Water-64

The Pest House-70

"Never Trust a Privateer"-74

The Timber Ship-78

The Bridge-80

"The Pocahontas"-84

Parting the Waters | Moses Pettingell-88

To Move a Lighthouse-92

The Mystery of the Missing Matriarch-96

Island Resort-100

The Jetties-102

The Wreck-106

The Gunners -110

Rum Runners-116

Rachel Carson-118

The Sea Haven Polio Camp-122

The Groins-128

The Paper-132

Sewer Lines on the Seashore-136

Geri's Loss-140

The Storm -144

The Pacific Legal Foundation-148

Academic Politics-152

The President's Day Blizzard -156

You Cant Sue the Atlantic Ocean-160

The Seawall Returns -164

North Jetty-168

The Matriarch's Message-170

The Last Folly?-174

A Golden Opportunity-176

Acknowledgements

I have mentioned most of the people who have helped me doing research for this book. But I would like to thank Chris Hein from the Virginia Institute of Marine Sciences and Dennis Hubbard from Oberlin College for reading the manuscript and particularly thank Becky Coburn and Jill Buchanan for their creative work designing this and former books. Josh Marshall has done a superb job marketing my books and Ethan Cohen has been fun to work with both in the field piloting his drone and in the studio editing down our monthly beach reports. I would also like to thank Jim Fenton for proving us with one of his spectacular landscape shots for the cover of this book.

7.05.16

For My Beloved Husband —
A fossil lover and — or
Islander forever + ever. So is my
love and my devotion to you.
Your Wife,
Lisee Anne

Fact and Fiction

Though it is based on fact, this book also uses some fictitious characters and dialogue to fully tell the story of Plum Island. For those who care about such things, this section will help explain which is which.

The Prologue is based on people's memories of the Blizzards of 2015.

Mica and Schist is based on our current understanding of plate tectonics,

The Glacier is based on the geology of the most recent Ice Age.

The Matriarch is based on a real mammoth that was unearthed on Plum Island in 1879. Her thoughts, emotions and actions are based on insights gleaned from animal behavior and the archeology of the Bull Brook site in Ipswich, Massachusetts.

The Sand is based on our current understanding of paleoclimatology and coastal processes.

The Marsh is based on estuarine ecology.

The Match is based on social anthropology and excavations on Plum Island and in the Merrimack River area.

Rips and Runnels is based on coastal geology.

Captain John Smith is based on our present understanding of Smith's interactions with Pocahontas, John Rolfe and the Powhaten. Snipes and his conversation with Captain Smith are fictitious.

The New England Prospect is based on the writings of William Wood and our understanding of how the mouth of the Merrimack River shifting north in the 1600's.

Wicked Walter Walis is a fictitious character based real people who cut salt marsh hay in Essex County and on Cape Cod.

Grape Island is based on writings about Elizabeth Perkins and Miss Jewett who lived on Plum Island. Their dialogue is fictitious. Nathaniel Pulsifer is based on himself.

The Blizzard of 1717 is based on writings about real events. The dialogue is fictitious but severe climatological conditions cold did deflate the Patriots' footballs in 2015.

Wits and Water is based on writings about the colonial economy. The Emerson family members were all real.

The Pest House is based on the real pest house on Plum Island. But Captain Sanderson is a gastroenterologist in London and Mayo Johnson is the former head of Surgery at the Beverly Hospital. I know he has good technique because as a kid I watched him hone his skills filleting flounder on Cape Cod.

Carl Soderland grew up in Woods Hole and is a general practitioner at the Lahey Clinic in Ipswich. None of these present day doctors had anything to do with the Plum Island Pest house but it would have been run with compassion and efficiency if they had. George Washington and Thomas Manning were real people.

The Timber Ship is based on Merrill Levi's 4 timber ships that were built in Newburyport. Geoffrey is based on a noted Newburyport fisherman and Michael is based on a modern day entrepreneur. The conversations between the two are fictitious.

Never Trust a Privateer is based on a real incident that occurred off Plum Island. Offin Boardman, Gideon Woodell and Cutting Lunt were all real people. Their conversation is fictitious.

The Bridge is based on real incidents. Benjamin Clifford was the real manager of the Plum Island Hotel. His conversation with his guest is fictitious.

Fact and Fiction

Pocahontas is based on the real wreck but the conversations in Lydia and Cutting's bedroom and at the wreck are fictitious.

Moses Pettingell is based on real people but their conversations were made up.

To Move a Lighthouse is based on real events but the people and their conversations are fictitious.

The Mystery of the Missing Matriarch is based on true events. The three gentlemen on the beach were real people and Louis Agassiz taught at Harvard but their conversations are fictitious

Island Resort is based on true events. Michael Simpson and the Pettengills were real but the Duggans and their conversation are fictitious.

The Jetties is based on real events. General Humphreys, James Eads and Captain Gagner were all real people and they conversation was as reported in the paper at the time.

The Wreck is based on real events on Plum Island and Cape Cod. Frank Stevens was the real Captain of the Life Savings Station. Mabel the horse was real but I gave her a pseudonym so she wouldn't take issue with anything I wrote. Watson was also fictitious but he sounded like he would make a good sidekick.

Rum Runners is based on real incidents but the names have been changed to protect the innocent-- and guilty.

Rachel Carson is based on her writings and her illustrator's memories of their wonderful summer spent on Plum Island.

The Sea Haven Polio Camp is based on the real camp and the American Chiropractic Institute's fight against the Salk vaccine.

The Groins is based on the real events that led to the construction of groins on Plum Island in the early Sixties.

The Paper is based on Dennis Hubbard's writings and personal recollections of the times.

Sewer Lines on the Seashore is based on the real events that led to the burial of sewer lines on Plum Island and their failure in 2015.

Geri's Loss is based on articles about the Thanksgiving Day loss of the Buzzotta house in 2008.

The Storm is based on eyewitness reports of the loss of 6 homes during a March storm in 2013.

The Aftermath is based on people's memories of the events that took place after the 2013 March storm.

Academic Politics is based on the real events surrounding the disappearance of Dennis Hubbard's paper in the scientific literature to be replaced by erroneous local knowledge,

The President's Day Blizzard is based on people's memories and television reports of real events involving the storm.

The Meeting is based on real events that occurred in the Massachusetts Erosion Board's meeting to collect public comments.

The Seawall is based on real events surrounding the illegal repair of Plum Island's illegal seawall.

The Matriarch's Message is based on the real events surrounding the reemergence of the woolly mammoth in, of all places, a newspaper article in a New York newspaper.

The Postscript is based on the real events surrounding the publication of a paper that predicted the sea would rise twice as fast and twice as high as originally thought and a bill that would provide $20 Million Dollars to buyout Plum Island homes threatened by erosion.

Prologue
The Blizzards of 2015
January 26, 2015

The Presidents Day Blizzard whacked New England just weeks after earlier blizzards had buried it under 5 feet of snow. This one packed hurricane force winds, whiteout conditions and 24 more inches of snow, sometimes drifting to 20 feet.

The Boston papers had already been running daily graphics comparing the height of the snow to Boston's favorite athletes. Party boy Rob Gronkowski was already buried under 6 feet 6 inches, and the Celtics' center Kelly Olnyk was next in line, at 7 feet even. They would all be buried when New England surpassed 10 feet in March.

But this storm was no joking matter. For the first time in all their years on the island, Plum Islanders were truly frightened. Even people whose homes were on pilings could feel them shudder with the impact of every wave. They couldn't hear each other talk over the sound of the wind and their electricity kept blinking on and off, so most went to bed early.

The next morning the wind was still howling, and 12-foot snowdrifts blocked everyone's front doors. One woman got a call from her father's assisted living facility. Apparently there had been a mistake and the drugstore had sent the wrong prescription. Could she go to the pharmacy and pick up the right medication?

She was met at the causeway by a group of local policemen. They explained there were severe whiteout conditions and she would have to wait for them to assemble a convoy of twenty cars so they could escort her off the island. They waited another 45 minutes for it to get dark so the drivers could see each other's lights, then they crept through the snow and wind only inches from the car ahead.

It had taken her most of the day to accomplish this simple task, but she had nothing but praise for the local policemen who had come up with the idea of the convoys.

Other people criticized the police for not telling them that if they drove off the island they probably wouldn't be able to drive back home. But the National Guard was also on hand in case anyone had to be relocated. These were life-threatening conditions.

20 Foot Snowdrifts on Plum Island-January 2015

But the most disheartening conditions presented themselves the day after the storm. Jayne Peng returned home to discover that her basement had filled with raw sewage. The same was true for at least 30 other people on both ends of the island and hundreds more were told not to take showers, flush their toilets or run any water. But that was OK, they could use a pot or the facilities at the Salvation Army's Building five miles away!

Sewage could even back up into empty summer homes, making them overrun with fungus. Imagine what it would be like when their owners returned to open up their houses in June.

Several families moved off the island to stay with relatives until the sewer system could be fixed, and half a dozen people signed up to be relocated through FEMA. It was like a mini-Katrina.

Prologue
THE BLIZZARDS OF 2015

Newburyport officials said that the reason the sewer system failed is that it had lost its vacuum, implying that homeowners had not kept their so-called candy cane vents free of snow. But the explanation didn't really make sense. Some of the homes whose vents were clear were also filled with raw sewage and some of the homes with clogged vents were doing just fine.

One of the affected areas was suspiciously close to where the ocean had washed over the streets where the sewer lines were buried. Had it undermined one of the mains causing a break? Had the rising tide caused groundwater to infiltrate the pipes?

But the bigger question was, why had the towns buried water and sewer lines under a barrier beach in the first place? Everybody knew that barrier beaches had to be able to move to survive such storms. Why build such a finicky system in an area prone to New England weather?

The answer has to do with taxes, money, misinformation and greed. But the real way to understand what is happening on Plum Island today is to go back 450 million years.

William Sargent
PLUM ISLAND 4,000 YEARS ON A BARRIER BEACH

Chapter 1
Garnet, Mica, Quartz, and all that other Schist.
450 million years ago

Mica, Garnet and Quartz

The best way to see how Plum Island formed is to drive up to Franconia Notch, preferably on a nice hot summer's day, and float on your back in Echo Lake. The water closest to your head trickles out of the west side of the lake to become the Connecticut River. The water closest to your feet seeps out of the boggy east side of the lake to become the crystal clear waters of the Pemigewasset River, and later the Merrimack River, that flows into the Atlantic Ocean between Salisbury Beach and Plum Island.

To the west are the gem-studded flanks of Garnet Hill and to the east is the Palermo mine where 6 foot-long crystals of quartz and 4 foot slabs of mica glitter against bluish green tourmaline walls.

450 million years ago the area that was to become Garnet Hill was at the bottom of a shallow sea hundreds of miles south of the equator. Trilobites scuttled over dark volcanic muds and ten foot long Eurypterids slithered over the bottom, capturing the jawless fish of this mid-Ordovician Sea.

But two major plates of the earth's crust were converging to close what we now call the Atlantic Ocean. As the East Plate plunged under the West Plate it melted and great blobs of magma rose to become an island arc of active volcanoes, similar to what you see in the Caribbean today.

One of these tropical islands was to become Garnet Hill, which gives its name to the catalogue company that brands itself as a purveyor of nice warm Northern clothes. It would give the Garnet Hill marketing department fits if it ever got out that their totemic northern hill was really the remains of a tropical island that once flourished 600 miles south of the equator.

All that dark Ordovician mud was buried under tons of sediment and eventually lithified into the slate that you see sliding off the face of today's Vermont road cuts. The limestone slate is also the reason that Vermont grows cows and grass, while the granite soils of New Hampshire can only grow trees.

During the Arcadian Orogeny that slammed the African plate into North America, the Ordovician slate was buried under seven miles of sediment, then cooked as the ocean finally closed and mountains the height of the Himalayas rose as the continents crunched together.

The slate then metamorphosed into Littleton Schist, a hardy rock laced with seven-inch long needles of greenish yellow sillimanite. Don't tell New Hampshire's Chamber of Commerce, but their iconic Mount Washington is actually made out of schist. not granite.

Chapter 1
GARNET, MICA, QUARTZ, AND ALL THAT OTHER SCHIST.

---

200 million years ago this area was part of the supercontinent of Pangaea but it too was starting to rift apart to open the present day Atlantic Ocean. As it did so, magma squished up through faults in the overlying formations to form plutons of pink Conway granite. The granite was once used to make monumental buildings and now makes up yuppie counter tops. But just enough of it remains to form the bedrock of Cannon Mountain that looms high over Echo Lake.

A little further to the east another ovoid blob of magma melted up through our old friend the Littleton schist and started to cool. It cooled from 800 degrees down to 50 degrees Centigrade over a long, drawn out period.

It was this long slow cooling period that allowed the magma to grow the masses of crystalline quartz and the slabs of biotite and muscovite mica, along with crystal filled pods of triphylite, and even more exotic minerals like uranium and Palermonite, found nowhere else on earth. In fact, the United States army keeps the Palermo mine on it's list of mines containing strategic minerals because it still contains a greater variety of minerals than any other place on earth.

So what happened to all those mountains that were once the height of the Himalayas? They eroded back to their volcanic cores, exposing the garnet encrusted flanks of Garnet Hill, the pink granite cliffs of Cannon Mountain and the semi-precious gems of the Palermo mine.

The mountains finally had a chance to rest after all those multiple orogenies. So they lay there, often under the feet of lumbering Mesozoic dinosaurs, waiting for the earth's next epoch that would see their minerals being torn out of their protective bedrock and delivered to their next incarnation, as the shifting sands of Plum Island Beach.

Chapter 2
The Glacier
18,000 years ago

"Glacial Epochs are great things, but they are vague, vague"
~Mark Twain, Life on the Mississippi, 1883.

Eighteen thousand years ago the area we call Plum Island was buried under half a mile of ice, the coast was three miles further east, and the sea level was 250 feet lower.

The glacier that started this process had expanded out of Canada and through Franconia Notch at the rate of about two football field lengths every year. Echo Lake and Cannon Mountain had been buried under two miles of ice; even Mount Washington didn't pierce through the featureless expanse of snow and ice.

This, the last glaciation of our present day Ice Age, was initiated when the earth's tilt and wobble had both conspired to cool the planet. But the cooling had already started by the creation of two earlier mountain ranges. The first were the Himalayas, whose mountains were so high they caused the earth's jet stream to bulge south, like it seems to be doing today.

More importantly, rain falling on the Himalayas formed carbonic acid, which incorporated atmospheric carbon dioxide into limestone, thus cooling the planet. Planktonic animals were also colonizing new shallow seas. When the plankton died their oily shells were eventually buried under several miles of sediment and cooked under all this pressure over 200 million years to form natural gas and oil. When we heat our homes and drive our cars, we are returning all that carbon dioxide back into the atmosphere where it is warming our planet once again.

The other important mountains were the string of volcanoes that rose only 20 million years ago to form the Isthmus of Panama. The Isthmus then cut off the warm equatorial currents that had formerly encircled the world. This deflected the currents north; creating warm, moisture laden clouds that dropped their moisture as snow on northern Canada.

When the snow grew to more than a mile thick, the ice beneath it turned plastic and started to flow southward. This initiated the Ice Age that was actually a series of glaciations and interglacial periods that continue today. Technically we are still in an interglacial period that would be naturally starting to cool if human activity was not overwhelming the natural cycle.

The last glacier to pass through New England was the Laurentian ice sheet. It pushed through Franconia Notch less than 80,000 years ago. Tons of ice tore fist-sized chunks of garnet off the subnivian flanks of Garnet Hill, gauged granite boulders from the sides of Cannon Mountain and plucked mica formations out of places like the Palermo mine.

Most of these rocks froze to the bottom of the glacier so when the glacier advanced, the rocks scraped deep gauges into the bedrock that you can still see today. Other rocks were pulverized into almost indestructible grains of silica, mica and garnet that today we call sand.

But then something changed. The Mallankovich cycles that determine the earth's tilt, axis and wobble came back into synch and the glacier started to melt more in the summer than it grew in the winter. This caused the glacier to melt back through Franconia Notch at the startling rate of about 4 football field lengths a year, more than enough to get Al Gore to reach for his handy Nikon.

The area we now call Plum Island was near the leading edge of this last glacier. As it advanced it carried its burden of sand, rocks and gravel forward. Then, as the climate warmed, the summers started to melt the leading edge of the glacier. But the glacier itself kept advancing, bringing more sand, rocks and gravel to the front.

Chapter 2
The Glacier

As the glacier melted, rivers of meltwater flowed out from underneath, depositing tons of pulverized sand in intricate, braided, deltaic patterns. Today you can still see long sinewy ridges of sand snaking through the New England countryside. One forms a sand mine on Rte. 93 in New Hampshire, another on Rte. 3 on the way to Cape Cod. Sand from these eskers is still being used to construct the streets and buildings of Boston. New York would also love to get its hands on our eskers but we would consider it larceny if not perverse.

As the glacier retreated it also reshaped great piles of glacial till into drumlins that look like grazing sheep with their heads all facing down in the same southeast direction. Today these drumlins rise as gentle islands above the surrounding marsh.

A string of these drumlins formed anchors that would one day hold Plum Island in place. But one of these drumlins was also the scene of a Paleolithic murder that would form the next epoch of our planet's ongoing story.

Chapter 3
The Matriarch
13,000 years ago

The Matriarch, by Kristina Lindborg

Thirteen thousand years ago a herd of woolly mammoth roamed east of what we now call Newbury. It was hot and the herd was plagued with flies and encumbered by fur evolved for a far colder climate.

The Matriarch of the herd had seen many changes in her fifty years. Every summer this thin strand of sand between the ocean and the ice had grown larger. The Wisconsin glacier, which had expanded as far south as Martha's Vineyard, had retreated back and was now in Maine.

The weight of the mile thick glacier had pushed this land 150 feet underwater and now it was rebounding. The coast was still 20 miles inland from where it would be in 2015. The matriarch had seen the land rise 9 inches in her lifetime. Every year she had witnessed the land gain another 36 feet. In time things would reverse and the land would retreat as the sea rose more than 200 feet in 3,000 years. Change was everywhere.

The land was still wet and glistening from the bluish gray mud that had been laid down when this area was still underwater. It had not had a chance to build up the marshes, lagoons and barrier beaches that are the hallmarks of a rising sea. Without these features, biological activity was low. There were few of the birds, fish and mollusks normally associated with a marshy environment.

Roving bands of Paleo-Indians, recently migrated into this area, still hunted in the more productive inland areas. As long as the herd stayed on the tundra-covered coast it would be safe from the packs of strange upright humans that didn't elicit fear because they were so unlike other predators.

Water was all over this boggy landscape. It cascaded off the face of stranded glaciers, pooled up in deep kettle ponds and collected in glacial lakes that were a hundred miles long.

The water nourished the tundra grasslands and swelled the rivers and streams that flowed toward the Atlantic. One of these was today's Parker River, that had started as meltwater on the surface of the glacier, then plunged down through a crevasse to excavate a channel for itself underneath the glacier.

But now the matriarch was faced with a dilemma. She knew it was safer to stay near the open coast but the herd was hot and getting deeply annoyed with all the flies and mosquitoes. Her own calf was running under her legs, flapping his ears and trumpeting his irritation.

The matriarch was none too comfortable herself. The insects seemed worse than anything she could remember. Finally she made up her mind. She lifted her trunk and sniffed the air for the smell of water. She remembered a boggy pond her mother had led the herd to when her mother had been the herd's matriarch. The pond's soothing waters had relieved the herd then, perhaps they would again.

Chapter 3
The Matriarch

With a determined air the matriarch wheeled and the herd followed after her. Excitement rippled through the older animals. They knew the matriarch would deliver them from their miseries. She had done so before and would do so again. With much jostling and trumpeting the herd trotted off into the nascent conifer forest.

A few hours later the herd spotted the pond shimmering through the forest of spruce and sedge. The mammoths broke into a run and splashed happily into the pond's cooling waters. Soon they were lolling about spraying each other with their long prehensile trunks. The matriarch's calf cavorted rambunctiously around his mother before spotting a spruce log drifting quietly on the surface. While his mother attended to the herd he swam over to investigate.

Suddenly a great shout erupted from the shore. Six men rose from the log and started splashing furiously toward the infant. Other Paleo-Indians emerged from the sedges and launched more logs from the far shore. The female tried to calm the herd and lead it back toward land. She knew that on firm ground the mammoths could hold their own, but in the water and boggy mud they were slow and vulnerable.

The matriarch had lost track of her calf. Two of the logs had cut him off from the herd and the men were closing in from both sides. If the infant had kept swimming he might have broken through their ranks. Instead, he panicked and turned back toward his mother.

The first spear glanced off the calf's shoulder, but the second struck and drew blood. It spurted from the concave fluting of the finely honed point. This was a nine-inch long Clovis point specially designed to hunt mammoth.

Now the calf was bellowing in pain and mounting terror. The matriarch heard his calls and but could do nothing. Another spear found its mark and more blood spewed into the now pink waters. One of the logs drew close enough so the man in the front could start clubbing the calf with a stone adz.

Each hit drew more blood but it was the one that cracked the calf's skull that finally knocked him unconscious. A shudder passed through the infant's body and he started to swim in jerky circles. One of the warriors jumped onto the calf's back and started to hack great chunks of flesh off his exposed flanks. The man's added weight was too much. The infant's unconscious body slipped underwater and spiraled out of sight. On shore the matriarch bellowed in rage and sorrow.

But the humans were not done. They turned their attention to the matriarch. They knew that if they killed her, the rest of the herd would just mingle around in fear while the men killed them one by one.

When the massacre was finally over, the humans stuffed gravel into the mammoths' intestines and used them to tie the carcasses to stakes driven in the soft mud. The pond's icy waters would both preserve the meat and hide it from other predators. The hunters would be able to return for more than a year to find the flesh, edible if not particularly appealing.

But the humans hadn't counted on one thing. Gas from the female's huge decomposing body was enough to float her to the surface until a spring flood washed her out of the pond and into the grip of the Parker River. There her body bumped and dragged until it finally became lodged in the shallow waters beside our drumlin. In time it would be incorporated into Plum Island. But that would not happen for another 8,000 years, for Plum Island is but a young and transitory feature on the face of our ever-changing planet.

Chapter 4
The Sand
10,000 years ago

"A beach is a place where sand stops to rest for awhile before resuming its journey to someplace else."
~Dery Bennett, 1977

When the Wisconsin glacier pushed through Franconia Notch it had pulverized the garnets of Garnet Hill, the granites of Cannon Mountain and the semi-precious minerals of the Palermo mine area into almost indestructible grains of sand.

The sand collected in rivulets, trickled into streams and was swept down the Pemigewasset and the Merrimack Rivers toward the waiting Atlantic Ocean.

But the Atlantic had been going through its own changes. So much fresh water had flowed into the ocean from the glaciers that it had prevented the surface waters from plunging toward the ocean floor. Normally, cold saline-rich waters plunged down through the water column, initiating bottom currents that transported the world's cold waters toward the Equator for warming.

These thermohaline currents were also the engine that pulled the Gulf Stream north. Without the currents, the Gulf Stream slowed, causing the Younger Dryads period; a 1,300-year long blip in the climatological record that first cooled the earth then warmed it so rapidly that in less than 50 years the average world temperature rose seven degrees in the North and four degrees worldwide.

The climate must have become unbearably hot for the cold adapted tundra species like caribou and woolly mammoth. They probably started retreating to the tops of mountains during the summers like moose do today, perhaps the rapid warming also lead to their demise.

It is something to ponder as we face a world that seems to be warming at the rate of one tenth of a degree every decade. That modest rate is already causing the Gulf Stream to spin off warm core rings and the polar vortex to swoop down into the interior of North America, causing droughts in the West and record breaking blizzards in the East.

What would happen if human induced warming altered the Gulf Stream as much as it did during the Younger Dryads? Would it trigger another Ice Age, a runaway Greenhouse effect or a world fluctuating back and forth between the two extremes? Fortunately it is the heat of summers as much as the cold of winters that determines if glaciers are going to advance or retreat, so we are unlikely to go back into an Ice Age but the febrile swings of a Mad House world is a real possibility. But I digress.

About 10,000 years ago, the Atlantic Ocean was two and a half miles further east from where it is today but rising fast. The Merrimack was shooting all that quartz sand offshore where it probably built up a large delta of Paleolithic sand.

Chapter 4
THE SAND

But it is equally possible that the sand flowed into a large open estuary that was half a mile wider than today. If so it would have looked like the Outer Bank's Albemarle Sound with its wide, open waters and long, skinny, fast moving barrier beaches.

By 6,000 years ago, the sea level rise had decreased enough so that sand bars and more stable barrier beaches could form and migrate slowly landward. They did this by a process that geologists call rollover.

During storms, waves can tear as much as a hundred feet of sand off the front of a beach and wash it over the rest of the island. The island essentially rolls over itself like the tracks of a tank.

We had a recent example of this process when Hurricane Sandy tore sand off the fronts of New Jersey's barrier beaches and deposited it in the streets of places like Atlantic City. In essence the entire state rolled back 30 to 60 feet during that storm.

So, Proto-Plum Island migrated in this manner for the next thousand years ending up close to its present location. There it stalled as it ran into the former upland drumlins that slowed its retreat. The drumlins would also provide a reservoir of sand that waves would use to gradually elongate the island. Today that reservoir is running out of sand, jeopardizing the homes of a different species. But that is a story for another chapter.

For now, it is enough to know that one of these drumlins contained the mummified remains of our female mammoth, and that is the reason you can find the body of a 13,000-year-old mammoth in a beach that is only 4,000 years old.

Chapter 5
"Rips and Runnels"
4,000 years ago

"Ridges and Runnels"

The Merrimack River area is unique in that during deglaciation the ocean rushed in to cover the land for a few thousand years, then rushed out to uncover the land for a few thousand more years, then rushed back in to inundate many of the areas where the Paleo-Indians had hunted Woolly Mammoth and migrating caribou.

Some geologists say the weight of the glacier was like someone pushing down on the center of a pie. This would cause the pie crust to rise. Then, when the cook removed his hand, the center of the pie would rise back up but the crust would dip down a bit and then return to normal. Remind me to never have dinner in those geologists' houses.

After all these fluctuations the ocean continued to rise, but at slower pace. It finally washed over Jeffery's Ledge around 9,000 years ago, undoubtedly drowning any Paleo-Indians and caribou that couldn't get off the shrinking neck of glacial till. Today Jeffery's Ledge is a favorite spot for catching tuna and watching whales.

The drowning of Jeffery's Ledge had severe consequences for proto Plum Island. The delta of Paleolithic sand had to then contend with the easterly winds that prevail along this coast. Without the interference of the old island, waves from the southeast could now pound directly on the Paleolithic delta. This initiated a period of beach building and migration that helped form proto Plum Island and continues today.

From the beginning these waves started to reshape Plum Island. They crashed against the shoreline and piled up against the barrier beaches. The water had to go somewhere, so it formed longshore currents that flowed in what today's fishermen call "rips and runnels" that parallel the beach, the geological term is "ridges and runnels ".

These longshore currents were the same ones that carry swimmers lazily along the shore today. But when such currents encounter a break in the offshore sandbars they turn directly out to sea, as the ill-named but treacherous rip tides that have swept countless swimmers to their deaths.

But the longshore currents also played a more prosaic role. They worked in collusion with waves to transport sand along the shore. Each wave would stir up sediments from off the beach, then support it in the water column as the longshore currents swept it parallel to the shore.

Then the next wave would redeposit the sediment back on the beach as a delicate filigree of sand grains. But each grain would have made an elegant little loop as it made its way downstream from where it had lain before. Even during the calmest days of summer, the beach moves past sunbathers who are blissfully unaware of this migration occurring right in front of their naked toes.

Chapter 5
"Rips and Runnels"

This process made Plum Island one of the most powerful sand transport systems on the planet. Both ends of such beaches can grow as much as 300 feet a year or almost a mile every ten years until they run out of sand in the middle. Astronauts can see this process from space. If we discovered such a system on another planet it would receive worldwide attention.

By 5,000 years ago the sea level was still 16 to 19 feet lower than today but it was rising 6 inches every fifty years, about what it is on Plum Island today. Plum Island Sound looked like its modern day counterpart but it was still several miles wider.

But there was one more complication. Even though the winds might be blowing in from the Northeast or Southeast, the waves tended to hit the beach directly from the East. This is because most of the waves that hit the coast came from miles away. When such a wave approaches the beach from an angle, its leading edge hits the bottom first and is slowed down by friction until the rest of the wave catches up and the entire wave hits the beach dead on from the east.

These prevailing waves would then hit the center of the island and create two sand cells of circulating currents. The northernmost cell would move sand from the center of the island north toward the Merrimack River and the southernmost cell would move sand from the center of the island south toward modern day Ipswich.

This made the island grow longer at both ends and narrower in the middle as it continued to move landward until it finally nestled into the drumlins that would slow its migration and anchor it more or less in place for the next four thousand years. So, after half a billion years we finally had modern day Plum Island. Whew!

Chapter 6
The Marsh
4,000 years ago

The Marsh

Around the time that the Egyptians were building their pyramids and Hammurabi was codifying his laws, a duck flew up from the South and a few marsh grass seeds dropped off his webbed feet into the shallow waters behind Plum Island. This was not an uncommon event. In the 1800s, Darwin counted the seeds of fourteen different species of plants caught in the mud on a duck's foot.

The seeds found the mixture of sand, mud, and salt water to their liking and started to sprout. They were nurtured by two crucial factors, the water was shallow and fresh groundwater seeped into the sound from the nearby island. *Spartina Alterniflora* seeds can sprout in two weeks when bathed in fresh water, but take over two months to sprout in undiluted salt water.

But Spartina had another trick up its sleeve. It had special organs for removing salt from its delicate internal tissues. If you look carefully at the base of a blade of Spartina you will see sparkling crystals of pure salt that have been excreted out of tiny, almost invisible stomata.

Using these advantages, the Spartina colonized the area by sending out long underground roots called rhizomes. The rhizomes sprouted up new shoots and sent down new taproots. The tangle of vegetation then trapped sediment to build up a foundation of squishy peat. This gradual buildup of peat allowed the marsh to keep pace with the rising ocean. Today the peat in the center of the Plum Island marsh is close to 18 feet deep, which indicates how much the sea has risen over the last 4,000 years.

As the peat expanded it started to prevent the high tides from inundating the upper marsh until the *Spartina alterniflora* was finally replaced by its less salt tolerant cousin *Spartina patens*. A new habitat had been created; lush green meadows, and dense stands of salt marsh hay.

The marsh was also laced with creeks that delivered nutrients with every high tide. This free fertilizer made the marsh ten times more productive than the most productive wheat field. The marsh was also dotted with salt-water ponds called pannes.

The pannes were formed when a patch of marsh died. The process happened when a high tide left a mat of eelgrass to die on the marsh, or a floe of ice grounded on the marsh and melted, dropping the burden of sediment frozen to its undersides. Both events would have cut off light to the Spartina, and without light the plants quickly died.

This made the Spartina roots start to rot, which formed a slight depression on the marsh. But it was deep enough to hold rainwater, which hastened the death of more marsh grass because Spartina can't survive in standing water.

Chapter 6
The Marsh

Like tooth decay, the mixture of fresh water and rotting vegetation bored a cavity into the marsh that then filled with mud and sometimes reached all the way down through the peat to the underlying sand. The pannes also filled with Ruppia weed, succulent water plants so favored by dabbling ducks and geese that its common name is wigeongrass.

It was these ducks and geese that would attract another species of two-legged creatures to Plum Island's productive marsh and beckoning beaches.

Chapter 7
The Match
November 22, 1606

The autumn sun rose slowly out of the Atlantic Ocean to shine on the tidy fields of the Pentucket village. The tribe was at the peak of its influence. Food was plentiful, trade flourished, and the sachem had kept the nation out of war for as long as anyone could remember.

A young warrior stepped out of his long house to greet the newborn day. He was a tall quiet man with clear, bronzed skin. At his feet lay a pile of clamshells, the remains of last night's feast. Attaquin swept the shells into a reed basket, walked to the marsh edge and dumped them there before reclining on a pile of deerskin blankets. He enjoyed sitting in the sun before his village was fully awake.

He looked out over the fields covered with the stubble of last summer's corn. The harvest had been good. Beyond the fields were open woods. Every year the warriors burned the underbrush to keep the forest open so they could hunt rabbits, deer and turkey in the tamed landscape.

The Merrimack River slid by the village into Pentucket Sound, whose protected waters provided the Pentuckets with the clams, striped bass and sturgeon that made up the bulk of their nutritious diet.

Scores of other villages hugged these shores that would someday be called Newburyport and Amesbury. Now it was simply a loose confederation of villages within the Pentucket nation.

As Attaquin walked back from the salt marsh, a runner from the Agawams was just arriving.

"Attaquin can you join us. Some whales have come ashore. Is your dugout ready?"

This was the moment Attaquin had been waiting for. For many months he had carefully hammered and chipped his knives. Now they were thirteen inches long. They looked like fine ceremonial objects but Attaquin had other plans for his finely wrought tools.

Attaquin and his brother were known to be the best fishermen in the village. They had grown up spearing sturgeon and salmon in the Merrimack, but after meeting some Wampanoags they had decided to go after bigger fish.

First they traveled up river to find a tall thick pine tree. They felled it and spent many moons using adzes and small fires to laboriously sculpt out its interior. When they were finely done, the brothers had a dugout that was long, wide and sturdy enough to paddle far out into the Atlantic where swordfish slumbered on the surface.

Attaquin and his brother made an efficient fishing team. Attaquin would stand in the bow, giving hand signals so Uncatena could quietly paddle up behind the unsuspecting fish. The two men would then hold their breath as Attaquin quietly slipped his spear into an atl atl to help it fly further. Then Attaquin would thrust the spear deeply into the muscular back of one of the somnolent giants. When the line went slack, Attaquin would tie deer bladders to the line so the fish had to fight to stay underwater. Eventually the fish would tire enough so that Attaquin could deliver the coup de grace. It was a long paddle home, and a long way from spearing sturgeon on the river.

The entire village would assemble when the brothers returned. Attaquin would clean the fish, remove any parasites and Uncatena would distribute pieces to every member of the band. He did this skillfully, garnering praise and political obligations. The sachem would thank the brothers, the elders would praise them, and all the young women would vie to catch their attention.

Now this rare discovery of whales would give Attaquin and Uncatena another chance to provision the village. With rising excitement Attaquin wrapped his long knives in deerskin, packed them carefully in the dugout's bow and pushed it into Pentucket Sound.

Chapter 7
The Match

Soon dugouts from other villages joined the brothers on the river. The Pentuckets paddled together into the sound and hauled their dugouts over the island to the broad waters of the Atlantic Ocean. A mile down the beach they could see the dark forms of the whales stranded on the sand flats. Each carcass was surrounded by half a dozen Agawam warriors hacking at the whales' tough skin. As they approached Attaquin called out.

"I am Attaquin of the Pentuckets. We have come for our share of the whales."

"What took you so long Attaquin? The whales are almost gone. Yours are the scrawny ones at the end of the beach." It was Pashto, Attaquin's old friend from the Agawams.

"How are you Pashto? Don't tell me you are in charge of the whales, we'll never get any. Is your sister here to see how real warriors butcher a whale?'

"Nananatuck is waiting. She plans to dance with you after the whales are butchered. But by that time you will have lost your loincloth to some real warriors. Did you bring your sticks? Tonight we start the first ball game, Agawams against the Pentuckets. Winners get the girls, losers lose their loincloths."

With such bantering the Pentuckets and the Agawams paddled down the beach toward the whales. Soon they were stripping great slabs of muscle-rich meat off the whales' bloody carcasses. Warriors from other villages wandered over to admire Attaquin's long knives that cut so deeply though the thick layers of blubber and meat. Attaquin could dress out two whales to their one.

"How much do you want for one of your knives Attaquin?"

"More than you can afford, Pashto!"

"Slow Duck here has a knife he traded with a white fisherman who sailed into our village an the last full moon. We don't like those white devils, but they keep coming back."

"You were probably too polite to them."

"Yes we made a big mistake. We showed them what good food we have. Now they won't leave us alone. Have you ever eaten the food Englishmen eat? I wouldn't feed it to a wolf. We ate some of the white man's and now look at us. We are tired and weak and Slow Duck has red bumps all over his body.

After the butchering was over, the villages separated into two teams. The warriors laughed and jostled each other as they battled with sticks and their bodies to dribble a small ball up and down several miles of beach. The games continued for four days. Afterwards the warriors hung clothes, spears and wampum on a driftwood arbor and rolled pieces of deer antler to gamble on each other's wares. The point of the contest was to bet, laugh, and swear loudly at the outcome. On the last night the villagers danced and Attaquin slept with Pashto's playful sister Nananatuck.

The following morning Attaquin and the rest of the defeated Pentatucks climbed into their dugouts to paddle back to their village. Nananatuck and her friends lined up at the top of the dunes to laugh at their new boyfriends as they paddled away naked.

Still the Pentuckets congratulated themselves. It had been a good clean beach battle, although perhaps too clean and not quite good enough. They vowed that next year it would be the Agawams who would lose their loincloths.

After several hours of heavy paddling, Attaquin and his brother returned to their village. They distributed the whale meat to the sachem and waiting villagers, and repacked their dugout with spears and Attaquin's new flensing knives. They covered the dugout with deerskins and buried it under earth and cedar boughs. It would be waiting for them packed and ready to go when the sturgeon returned in the spring.

Chapter 7
The Match

Little did the Pentuckets know that the hepatitis that Slow Duck had picked up from the white man's knife would race through the village and nobody would be alive to unearth the dugout canoe the following spring.

Chapter 8
Captain John Smith
April, 1614

Captain John Smith was in foul humor — per usual.

"Snipes, assemble a detail on deck and have them move sharply!"

"Yes sir, and how was your breakfast, sir?"

"My breakfast? Who gives a damn about my breakfast? I didn't like the looks of that smoke rising beyond the far hill. "

"But Champlain described these Indians as peace loving people, sir."

"Damn that Frenchman's eyes. It was he who had those Monomoyicks killed, all because his sailors stayed ashore so they could finish baking their bloody French bread on Mallebarre. Remind me to change Mallabarre to Cape Cod in my next book, Snipes!"

"Yes sir, Captain Smith!"

"And Snipes, make sure the men bring their muskets. Dismissed!"

"Vainglorious bastard," muttered Snipes as he backed out of the captain's quarters.

"Ah, it feels good to be back in charge," Smith thought to himself. His had been an adventuresome life. He had been born in Lincolnshire and from that day forward had determined to avoid his father's humdrum world, to seek fame, honor and glory as a mercenary soldier.

But the life of a mercenary had its drawbacks. On his first assignment he had been captured and sold into slavery in Turkey. But another one of Smith's traits was that for some reason women always seemed to want to get him out of trouble. His kind-hearted mistress couldn't stand to see the robust young man in slavery so she arranged to have him work on her brother's farm.

Smith returned the favor by killing the brother and escaping back to England where he signed up with Bartholomew Gosnold commanding colonists heading to the Virginia Colony. But even aboard a ship he could get in trouble, and was accused of mutiny and locked into his cabin for the duration of the voyage.

Once on land, John helped overthrow the rule of Edward Wingfield and was rewarded by being put in charge of leading an expedition, up the Chickahominy River, to deal with the Powhaten Indians who also claimed the colony's land.

Things didn't go well and Smith was imprisoned once again. But the Indians treated him well, so well in fact that chief Powhaten held a ceremony to ritually kill the white man so he could be reborn as his son in the Powhaten nation.

But the chief's 12-year-old daughter Pocahontas, "the mischievous one," was so smitten with the charismatic white man that she broke up the ceremony. John figured he could embellish the story a bit, but would wait for 17 years until Pocahontas was safely dead and could not refute his apocryphal version of the ritual.

For her reward, Pocahontas was raped and carried back to the English colony where John Rolfe fell in love with the still captivating young woman.

Smith didn't know it at the time, but just as he was about to set foot on Plum Island, Pocahontas was marrying Rolfe, believing that Smith had been killed years before in Virginia.

Chapter 8
Captain John Smith

But enough of such musings, Captain John Smith was a man of action and had an island to explore. In his book entitled *Explorations of New England* he would later write:

"On the east is an isle of two or three leagues in length; the one halfe plaine marish ground fit for pasteur, or salt ponds, with many fair high groves of Mulberrie trees and gardens; and there are also Oaks, Pines, Walnuts and other wood to make this place an excellent habitation, being a good and a safe harbor."

It was said that he described Plum Island so well, that the Pilgrims shivering in their hovels in Plymouth considered pulling up stakes and moving north to this perfect place.

"Explorations" was published to great success in 1616. And it was during that year that Pocahontas, now the respectable English lady Mrs. John Rolfe, saw Smith for the first time after she had left Virginia. Smith had never bothered to tell her that he was still alive and he had never tried to assuage the bad blood that had developed between the Powhaten and the English.

Pocahontas turned on her heel and had to leave the theater in order to calm herself. The following year she decided to sail back to Virginia but she was so sick she had to be taken off the ship. She died and was buried at the Thames port of Gravesend, at the age of 21.

But John had one more treachery up his sleeve. While exploring the New England coast, one of his lieutenants had captured several Massachusetts Indians, one of whom we know as Squanto, the patron saint of Thanksgiving, who saved the Plymouth Plantation by teaching the Pilgrims how to hunt, fish and plant corn, beans and squash.

Pocahontas had taught the Virginians how to cure the tobacco John Rolfe had brought back from the year when he had been shipwrecked on Bermuda. That, along with the peace of Pocahontas, had made the Virginia plantation profitable just before the investors were about to pull the plug on the whole operation.

So what did these two Native Americans who did the most to save the British colonies have in common? They had both been screwed by Captain John Smith who had gone on to fame and fortune extolling the virtues of colonizing the wilds of New England, that had been tamed and landscaped by the Algonquin tribes several centuries before.

Chapter 9
"The New England Prospect"
1634

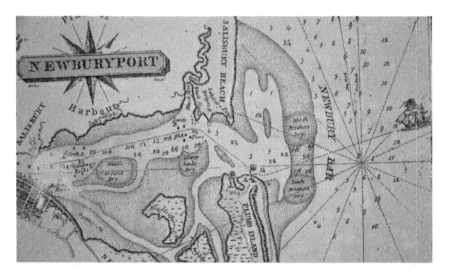

1634 Chart of the Mouth of the Merrimack River

The geological record shows that the ends of barrier beach islands like Plum Island are fast growing but unstable. We saw how the southern end of Plum Island grew several miles forcing the mouth of the Parker River south to enter the Atlantic in Ipswich rather than through its former location in Newbury.

That process took thousands of years. But the same was happening on the northern end of Plum Island, witnessed over the course of hundreds of years.

Captain John Smith got all the glory for exploring the New England Coast, mostly because of tall tales told about his shenanigans with Pocahontas. But to many minds, William Wood was the far better observer of the natural scene.

In 1634, Wood wrote that, "The Merrimack is a gallant river whose mouth was over a mile wide, but it has two rather 'indifferent' channels separated by an island of sand."

It was the shallowness of these indifferent channels that convinced the early settlers to establish Ipswich on the Ipswich River and Newbury on the Quascacuguen River because, both had such "fair channels" that 50 to 60 ton vessels could navigate them without incident.

Several years later, Newbury changed the name of the Quascacuguen River to the Parker River, a little boring, but far easier to spell.

Almost immediately, the residents of Newbury started to feel their oats and expand north to the shores of the Merrimack for fishing and trade. They also made the argument that, "all of Plum Island deserves to be theirs because... in right it belongs to us and without which we see no way to continue to subsist."

But the General Court didn't buy Newbury's rather circular reasoning and divided Plum Island between Ipswich, Rowley and Newbury in 1649. The year before Newbury had laid out a new town on the Merrimack River and commenced to bicker over whether to move the meeting-house north toward the new town that we now call Newburyport.

By the mid 1600s people were starting to build small vessels to trade salt cod, timber and beef for sugar, tobacco and indigo from the Carolinas and the West Indies. The trade became so lucrative that in 1656 Captain White built a wharf at the foot of Fish Street, now called State Street and the inhabitants of the new town started to earn money from maritime occupations while the people of old Newbury continued to grow turnips and argue about details of an ecclesiastic nature.

It was also during this time that people first started to pack their boats and sail to Plum Island in order to fish, dig clams and enjoy the sun on hot summer's days, between rounds of cutting salt marsh hay and caring for the cattle that ran free on Newbury's end of the island, much to the annoyance of Ipswich and Rowley farmers who kept theirs properly fenced in.

Chapter 9
"The New England Prospect"

But the morphology of the island was changing as well. The two channels and the sandy island in the mouth of the Merrimack that William Wood had described in 1634 had deepened to become a single more navigable channel. Later, 600 feet of the beach would be swept away and vessels with as much as a 7 foot draft could sail all the way around Plum Island at dead low tide. Undoubtedly this proved to be a boon to those who wished to avoid detection by the British Royal Navy to say nothing of their own later tax collectors.

Then in the 1830s sand closed off the mouth of the river entirely, so it formed two channels with an island in the middle, the exact same pattern that William Wood had described in 1634. In doing so it also lopped half a mile of sand off Salisbury Beach and gave it to Newbury, or more precisely to the sand miner, Moses Pettingell, who then owned the north end of Plum Island.

Chapter 10
Wicked Walis
August 10, 1640

Delicate puffs of sea lavender dotted the shore and the heads of salt marsh hay were ripe with seed. Robert Manning figured he had just enough time to make one more cutting before the salt hay collapsed into long windrows, that always reminded him of cowlicks on a bearskin rug.

It had been twenty-four years since Captain John Smith had extolled the virtues of this marsh and he had been right. Being able to graze their livestock on the marsh had allowed the early settlers to make it through their first year in Ipswich.

The peat in this young Spartina marsh was not very deep. In the older areas of Agawam Bay it was close to 15 feet but was adding an additional foot every century as the sea level rose.

Peat was not so important, but hay was the mainstay of the New England economy. You needed it for food, transportation, even housing. All of Manning's neighbors packed Spartina straw against their foundations to insulate their homes.

Of course the best thing about salt marsh hay was that it was free. You didn't have to clear the land, or lime and fertilize the soil. The tides did that twice a day for nothing. Upland farmers also preferred to feed salt marsh hay to their cows because it didn't contain the weed seeds that could pass through the cows' multiple guts and come out in their manure to sprout in the farmer's fields of English hay.

For the last few years Manning had cut twice as much salt marsh hay as English hay. He was even considering transporting his excess to Boston to sell in Hay Market Square.

While Manning hitched up his oxen to the hay wagon, Walter Walis gathered in the hogs. The partners had been granted the right to care for fourscore hogs on the island as long as they stayed with them constantly, and returned them to their owners for slaughter in the fall. They would be paid 4 pence a hog for their labors at the beginning of the summer, 2 pence at mid-summer and 2 pence for as many hogs as had survived by harvest time.

Walis was certainly good with the animals and had a fine strong back, but Manning thought the young buck liked his liquor far too much for comfort. If only he would show up for work once in awhile and not get into quite so many fights over the tavern women. The other men had taken to calling him "Wicked Walis" but the young man seemed to like their sobriquet. This was not a good sign in Manning's book.

The farm hands spread out into long lines and swung their scythes in slow unison, as they cut their way across the marsh. Walter stayed with the horses on the hard-packed sand at the edge of the marsh.

In the 1800s George Randall from Newbury would put large wooden horseshoes that looked like snowshoes on his horses hooves so they could walk across the marsh without getting stuck in the soft peat. Before that, scores of horses and oxen had drowned when they became mired in the many pannes that pockmarked the marsh.

When Manning's team was done cutting, the men drove long slender trunks of black locust into the marsh in large thirty-foot wide circles. The peat would prevent the locust poles from rotting, so these staddles could be used to dry the salt marsh hay for years to come.

Manning was thinking of building a causeway through the marsh so his oxen wouldn't sink in the mud. It would take an immense amount a labor. They would have to remove tons of boulders from the Cross Bar drumlin, load them into horse drawn carts, then use them to build the causeway. Walis said he was damned if he would heft hundred pound boulders in ninety-degree weather while fighting off midges, greenheads and mosquitoes.

Chapter 10
WICKED WALIS

The causeways would become extraordinary pieces of engineering. Extra high tides would cover them with several feet of water, but otherwise they would remain unscathed for centuries to come. Perhaps this is why the North Shore's General Patton knew he could drive his tanks over the old Roman roads without getting stuck in the soggy Italian countryside during World War II.

By eight o'clock the sun was finally starting to sink behind Town Farm Road. Walis retrieved his rifle from Manning's shallop and drove the horses to a temporary barn they had built on one of the Agawams' old shell middens. He was to spend the night protecting the horses and hogs from wolves while the rest of the farm hands returned to Ipswich in Manning's shallop.

It had been a long day. The men took one last draught of beer and packed their scythes into Manning's boat for the short sail to Greene's Point. They were happy the cutting was almost over. Tomorrow they would toss the hay so it would dry evenly, then stack it on the staddles until the marsh froze and they could pile it in hay carts to take back to Manning's farm. They had already loaded several long shallow gundalow barges full of black grass for winter fodder.

The following morning Manning sailed back across the sound, but Walis was not on the marsh. He was supposed to have driven the horses back up the beach to meet the men at their predetermined spot. Had he been attacked by wolves or swept off the beach during one of his late night swims? No, he had taken a boat across the Sound and spent the night in a pub while wolves had methodically killed most of the town's supply of hogs.

From that day forward people became obsessed with the exploits of Wicked Walis and his multiple descendants. If you asked someone on the North Shore how to tell the difference between a Walis with one "l" and a Wallis with two "l"s they would they would have a simple answer, the Walis with one "l" killed other people's hogs and the Wallis with two "l"s did not.

The exploits of the Ipswich men so upset the town fathers of Newbury that they used it as an excuse to try to wrest control of the entire island from Ipswich. The two towns squabbled for 11 years until the General Court finally granted two fifths of the island to Newbury, two fifths to Ipswich and a piddling one fifth of the island to the town of Rowley.

After several successful decades, North Shore farmers stopped the intense haying of Plum Island. By the 1800s most of the forests had been cut down for firewood anyway, so it was easier to simply cut English hay in your own field, closer to home, though the Colby family kept cutting marsh hay right up until 2015.

But in the far reaches of the marsh, Manning's slender trunks of black locust remained buried in the peat, and Wicked Walis' footprints remained buried under sand in the hard-packed mud. The sand had blown in and would preserve his prints like perfect fossils until the beach rolled over it once again and an autumn storm 300 years hence would wash away the sand and make it look like Wicked Walis had walked off the island only the night before.

Chapter 11
The Ladies of Grape Island
1680's

Elizabeth Perkins had always been too outspoken for the good people of Ipswich. As a young girl she used to talk back to her parents, but it was only when she accused her local minister of immoral acts that she came to the attention of the town fathers.

They found her guilty of being a "virulent, reproachful, and wicked-tongued woman," sentencing her to sit in the meeting house during services with a sign pinned to her bodice proclaiming in capital letters that she had reproached, "ministers, parents and relations." She was further ordered to be, "severely whipped on her naked body," but the elders were to be disappointed. She was able to come up with the three pounds to pay off the fine instead.

It was only when her husband also ran afoul of the law that Elizabeth decided.

~ Elizabeth Perkins ~

"Luke, we have to get out of this town. I can't stand these people!"

"Where are we to go? I have no decent prospects."

"But you have a strong back, Luke. You have always loved haying. We can rent Wainwrights' farm on Plum Island."

"How do you know he will rent to us?"

"I spoke to him yesterday on the town green. He likes you and has also had enough of the town fathers."

"Do you really think we could make a living out there all year long?"

"We wont know unless we try now will we?'

"Tis better than living beside these fools. How much is he asking?"

"Five pounds for the year."

"Let's take it. No questions asked!"

"You can scythe hay all summer and I shall plant a garden."

"Good man Wainwright would probably even ferry his cows over so we could care for them on Grape Island."

"Yes that's the spot. The deep creeks will keep the cows fenced in and those damn free ranging Newbury cattle out so they wont trample your turnips."

"And all the clams we can dig, and all the fish we can catch for the rest of the year."

The hard working young couple were soon accepted by the Pulsifers, Treadwells and Wainwrights who had farms on the island. Luke helped with their extra farm work and Elizabeth cared for their houses during the winter months.

~ Miss Jewett ~

Miss Jewett was a little frightened when she arrived on Grape Island in 1881.

"Thank you for the ride over Mr. Pulsifer."

"Not at all young lady, not at all."

"Do you know where Mr. Wainwright's house is? I'm to stay there all summer to teach the island children."

Chapter 11
THE LADIES OF GRAPE ISLAND

"Of course you are, and you will do a fine job too."

"Oh I'm not too sure of that. I don't know how I'll keep them concentrating on their books when this beautiful island beckons outside."

"Well I'll give you a hint. When the children start to get restive, just say you have to go to the outhouse and act astonished when you return and find the hands of the clock have been advanced by some of the taller students."

"Oh yes I'll just say, 'Oh my, doesn't time fly by fast on this island,' and let them out five minutes early!"

"They will love you."

"Perhaps you could take us out on your boat next August. It would be a small compensation to the children for having keep their minds on their books all month while the mainland kids are enjoying their vacations."

"That is a splendid idea. I'll see if Good Man Appleton will give us some fresh milk and I will bring my wife's ice cream maker. The children can churn it up on the way to the beach."

"Oh I think I will love teaching on this island, Mr. Pulsifer."

"Of course you will."

And she did. Miss Jewett taught in the old one-room schoolhouse on Grape Island until she retired 35 years later.

Chapter 12
The Storm
March 7, 1717

Wolves, Wolf Hollow, Ipswich, MA

Jack Richman stamped his feet to keep warm. It was already the coldest winter anyone could remember.

"Boy, bundle yourself up, we're going to sail to the island to get some firewood."

"Jack, are you sure? There's five feet of snow on the ground."

"Won't be any on the Sound. Besides, if we don't get any wood, we're sure to run out before this winter is over."

The Richmans didn't know it, but a series of volcanoes that had erupted in 1716, had cooled our planet and formed nuclei for snowflakes.

Sarah bundled Jack junior up in warm clothes and bid the two goodbye. She knew Jack could take care of himself but Jack junior was the apple of her eye and only 11 years old.

It was gray and overcast, but Jack figured they could make camp, load the boat and be back before the weather closed in on Sunday. But by noon the winds had shifted to the Northeast delivering a mix of snow, sleet and pelting rain.

"Let's cut short our wood collecting and get back in the boat, Jack."

Jack junior was only too happy to agree, but it was too late. The overloaded boat wallowed in the waves sweeping across the sound and the winds tore at their wet, heavy canvas sails. Their mast finally snapped and the two were blown onto the mainland marsh.

"We have to abandon ship here son. Think you can make it across the marsh into town. Mother should have some piping hot soup on the table."

"But my feet are getting cold."

"Nonsense boy, we'll be home in a few hours."

But the ice and snow made it difficult to walk and impossible to see the pannes hidden in the marsh. Richman fell through up to his waist several times. The last time was too much and Jack hugged his son and said he was sorry as they waited to slowly die.

By the time the townsmen were able to put together a search party, it was already far too late. The first major snowstorm had hit on March 1st, followed by another on the 4th and the worst on the 7th.

The snow would lighten for a few days between the storms but the sky never cleared until the next storm made it impossible to search the marsh. It was mid-march before the search crew could get out and find the frozen bodies of Richman and his son.

Chapter 12
THE STORM

In town, the Native Americans said they had never heard the elders speaking of such brutal storms. Most of the settlers' single story homes were so deeply buried that you couldn't even see their chimneys.

People got in and out of the larger homes by climbing through their second and sometimes third story windows. The snow lay ten feet deep over large expanses of land and drifted as high as twenty-foot where the winds had been most severe.

Hundreds of horses and cattle starved to death or were caught and frozen under drifts of blinding snow. Deer were unable to find food and were exhausted from trying to run through the tiring snow. One by one, wolves chased them down until over 90% of them and been killed and eaten.

The following Autumn, the deer reeves were given permission to check barns and homes to see that the remaining deer were not being over-hunted. Farmers discovered that their harvest of apples and pears was meager because the snow drifts had been so high that rabbits had been able to devour the buds of the branches twenty feet above the frozen ground.

New England would not be visited by such a series of blizzards until 2015. An Oklahoma climatologist calculated that the chances of so much snow falling in such a short time was over 26,700 to 1. When Boston passed 110 inches, one wag intoned, "Boston beats 26,700 year record. Hell freezes over. So cold, footballs deflate. Patriots win super bowl."

Chapter 13
Wits and Water
1731

Nathaniel Emerson's Domain

Nathaniel Emerson gazed out over his Plum Island domain. "Not bad for a lad from Ipswich," he thought to himself.

When the General Court divided up the land between Ipswich and Newbury it had been advertised as fine upland farmland. But it was neither fine, nor farmland. The soil was far too sandy and salty to grow much of anything besides salt marsh hay.

However, Nathanial had other plans. After he bought the southernmost drumlin on the island, he used its rocks to build a wharf and the trees growing on its bluff to build stages for drying codfish. He anchored his two fishing sloops in the creek behind the island where the Frenchman Gilshennon would later draw ocean water into wooden troughs to let the sun evaporate it into salt.

Salt was as important to the 18th century economy as hay. So much so, that English towns that ended with the suffix "wich" meant they were where salt works were located, so the original Ipswich had been where the salt works on the river Orwell in East Anglia stood. So, the East Anglia farmers who ended up in Essex County knew how to cut salt marsh hay and make salt. It was in their blood.

Nathanial started to sell salted fish to the West Indies where it was used as food that was both easy to store and inexpensive to feed to slaves. There was little way around it. If you wanted to earn money in the British Empire you often ended up supporting the trade in slaves.

If only Broster wasn't such a wastrel the Emersons could have kept Nathanial's fishing operation going, but Broster had squandered his father's money and the operation had been sold to Ralph Cross who had built up Cross Farm.

Such salt meadow farms were the other route to financial success in the New England colonies. They weren't as expensive and time consuming to start as an upland farm because you didn't have to clear the land first. Most farmers only had themselves, their families and if they were lucky a few temporary paid farmhands, so it was impossible to clear much land.

But land was cheap in the colonies. It was labor that was expensive, exactly the reverse of what the settlers had left behind in Jolly Olde England. In fact salt meadow farms were so important that by 1690 Essex County had more inhabitants than any other county in Massachusetts other than Boston that was also surrounded by marshes.

Then during the 1700s farmers expanded from having a few cows and horses for their own needs to raising hogs for money. The hogs were easy to maintain, unless they were tended by men like Wicked Walis. You could let the hogs loose to fend for themselves all summer then slaughter them in the fall for cash.

Chapter 13
WITS AND WATER

After making a little money from selling pigs, the farmers would move on to buy more cows. But the cows required more work and more farm hands. You had to build permanent barns and cut hay to feed the cows all winter.

But the small herds provided milk, cream, butter, cheese and eventually meat. More importantly cows provided farmers a way to earn money all year long.

Some farmers also found they could cut white oak and fashion it into barrel staves to be used in the West Indies for holding molasses. Other farmers starting using cow hides to make shoes to sell to visiting sailors. Later they would learn how to cut timber and float it down the Merrimack River to build fishing boats and eventually the beautiful McKay clippers for the China trade.

One enterprising Yankee even came up with the meshugena idea that you could make money by cutting blocks of ice out of New England ponds and selling them to plantation owners in the West Indies. Frederick Tudor's first boatload of ice melted on the docks in Martinique. Talk about liquefying your assets.

Bostonians started snickering about Tudor's folly, But several bankruptcies later his folly paid off and New England farmers started spending their winters cutting ice to sell to Tudor who would ship it to ports from Calcutta to the Caribbean.

Today you can still see the massive three foot thick walls of the ice warehouse on Tudor's wharf in Charlestown and you can still drive past innumerable "Silver Lakes" and "Crystal Ponds" once unprepossessing sloughs renamed to enhance their ice making image – relics of a New England Ice Age of decidedly more commercial nature than the one that built up the drumlins of Emerson's Plum Island.

But here's the rub. The West Indies had sugar while New England was stuck with ice. Yankee ingenuity or not, who wanted to be out cutting ice all winter so some fat cat plantation owner could sip cold drinks while slaves did all his work?

Of course places like Ipswich and Newbury were backwaters. The sugar plantations of the West Indies and the tobacco plantations that John Rolfe and Pocahontas made possible were the real moneymakers of the British Empire. In 1640 the New England colonies only had 40,000 inhabitants while the West Indies had 80,000 citizens not including their slaves.

By 1770 hundreds of New England towns, like Ipswich and Newbury, had their own distilleries for converting West Indian molasses into rum to fuel the lucrative sugar, rum and slave trade. Besides sugar was becoming more common in New England as tea was coming into vogue. No wonder New Englanders staged a tea party in Boston Harbor. But notice they dumped tea not sugar into the harbor. Sugar was too valuable; it could be traded for slaves.

So how could codfish, hay and ice really compete with sugar and tobacco? Who wanted to scratch a living out of New England's rocky soil instead of owning a plantation in the Caribbean?

The answer says a lot about economic opportunity and democracy. A few people could make a lot of money in the South but a lot of people could make a little money in the North. After all, who but a Yankee would think you could earn a living shipping ice to the Caribbean? Perhaps that is what historians meant when they said the way to make money in New England was through wits and water.

But maybe they should have added just a soupcon of chutzpah to make the adage ring true. Whatever it was, New Englanders were finally able to accrue enough capital to enjoy places like Plum Island as merely summer spots and not have to use them to make a living.

Chapter 13
WITS AND WATER

It was the basic fairness of the New England system that made it so successful and long lasting. Even today nations whose economies are based on a wide range of natural resources and distribute their wealth more equitably are more democratic than countries whose economies are based on a single natural resource like oil. And it all started on places like Plum Island where you could earn a good living if you could put up with hard work, heat, and those swarms of pesky mosquitoes.

Chapter 14
The Pest House
1769

"How goes the battle, Sanderson?"

"Not good, Captain. Two men are down with fever and covered with rash. Ship's doctor says its smallpox."

"Damn, have the watch raise the quarantine flag and lock the men in their cabins."

"How will we feed them, sir?"

"Have the steward pass them food under the cabin door and order the men to throw their wastes out the port holes."

"What about their blankets? They've soiled them."

"Keep those in their cabins as well. As soon as we reach Newburyport they will have to be buried. I understand there is a pest house on the island just before you reach the port."

"Will they send a boat out to meet us?"

"Hopefully so, and have the officers keep all the other men below decks during the transfer. I don't want anyone to come in contact with those lepers."

"They have smallpox, sir."

"I know they have smallpox. It was a turn of a phrase, Sanderson. Now get the hell out of here or I'll have you swab down the whole damn ship with vinegar after the men are gone."

"Yes sir, Captain sir."

On shore Carl Soderland saw the quarantine flag as the *Seahorse* rounded the point.

"Have Johnson and Mayo row out to meet the unfortunate souls and pull together a detail to bury their clothes. And be sure they bury them deeply enough this time. We don't want any pestilent blankets drifting into the harbor again."

"Anything else, sir?"

"Yes read the patients the rules and tell them that if they attempt to escape before they have recovered they can expect to be stoned by the townsmen. And make sure nothing is burned. Next spring we will have to build a fence to keep cows from eating grass where we buried the clothes."

Nobody quite knew why none of Soderland's staff didn't come down with smallpox themselves. Perhaps they had been exposed as children or picked up cowpox from milking cows.

Whatever it was, he only hired people who carried the telltale scars of the disease and carefully explained that they would be isolated all winter on the frigid island. But most were happy to have employment.

Johnson and Mayo rowed out to receive the sailors who screamed with agony every time the men pulled on the oars. Most of the Pest House patients came from England and had the minor form of the disease but the *Seahorse* had arrived from the West Indies and the two sailors had the hemorrhagic form of the disease, that the workers called black pox.

The pox had so shocked the sailors' immune systems that they could no longer make pus and they were covered with crusty pustules of *Variola Major*. Their blood had turned the color of coal and was seeping uncongealed beneath their skin and oozing out of their mouth and eyes.

Chapter 14
THE PEST HOUSE

The poor men stared at their saviors with wild, bloodshot eyes. They felt as if their insides were coming apart and in a sense they were. The pox was destroying their skin both inside and out. It had already liquefied the linings of their throats, stomachs and rectum.

When the boat reached shore one of the sailors lurched forward, and blood speckled with bits of liquefied intestine gushed out of his burning anus. Mayo and Johnson tried to hide their horror as they helped the men out of the boat.

But the fever was too much and one of the men threw off his clothes and ran into the icy cold waters of the Merrimack. Hopefully the waters provided the man with some small relief before they triggered the heart attack that ended his short, unhappy life.

The following day Johnson and Mayo carried the corpse up the beach and buried it along with the man's clothes.

Not too many years hence, Dr. Thomas Manning from Ipswich would ask his brother in England to ship him some cowpox lymph and use it to break Dr. Waterstone's monopoly on vaccinating people against smallpox, at $150 a pop.

It was far better than the variolation technique that George Washington had used to inoculate his troops. That had merely consisted of scraping the puss out of the pustule of someone with smallpox and scratching it into a healthy man's arm. The healthy man would then get the disease and if he was lucky it would be less virulent than the first man's disease and he would recover with lifetime immunity from the dreaded affliction.

After Washington made the decision to variolate his entire army it had to be done in absolute secrecy, often over the objections of men who didn't want to undergo the dangerous procedure. During the height of his variolation campaign, Washington only had 7,000 soldiers healthy enough to fight. If the British had attacked in March 1777, the war would have come to a rapid and unfortunate conclusion.

However, by the time the British discovered what had happened, over half of Washington's troops had recovered enough to continue the war. Washington would always say that his decision to variolate his troops had been as important as any strategy he had used on the battlefield.

But history has a way of remembering the deeds of humans, while forgetting the role of germs. By the end of the Revolution over 25,000 soldiers had died in combat but at least five times that many Americans had died from smallpox.

Of course all this was lost on the unfortunate men and women buried in the unmarked smallpox graveyard situated somewhere on the northern end of the desolate island.

Chapter 15
"Never Trust a Privateer"
January 15, 1776

During the Revolution, Newburyport had one of the largest fleets of privateers on the East Coast. Some of her vessels traveled far afield and boasted multiple cannons and swivel guns. Others operated closer to home and guile proved to be their most effective weapon.

Gideon Woodell: "Look, there is a ship tacking back and forth off Plum Island."

Johnson Lunt: "She looks British and lost I'd say."

"We better fetch Offin. He'll know what to do."

"And where might the good Captain be at this time of day?"

"Ye Olde Wolfe's Tavern, of course. Drunk or sober Captain Boardman will know what to do."

After a quick splash of water, Captain Boardman had a plan.

Offin Boardman: "Gideon you and Johnson get your whaleboats and plenty of pistols but keep them out of sight."

Soon the three whaleboats with seventeen men aboard were rowing out to meet the British ship.

"Ahoy the boat, ahoy the boat. From where do ye hail?"

Captain Bowie: "From London bound for Boston, sir."
"We are from Boston. Do you wish a pilot?"

"That we do, sir."

"Bring her around. We'll pilot you in."

"Excellent. Boson throw down the ladder!"

While the two gentlemen discussed the news of London, Boardman's men mounted the ship's gangway and quietly assembled on deck.

"Captain Bowie you are a kind gentleman indeed. So it is with regret that I have to inform you that you have been boarded by the Newburyport Irregulars. Strike your colors sir and I assure you nobody will be harmed."

"Why you are nothing but a bloody pirate!"

"On my word no. Not a pirate, a privateer, sir."

"You will pay for this Boardman!"

"Maybe, but not today!"

"Alright men strike the flag. We have no arms, resistance is futile."

"Good decision Captain. Now have your men land her at yonder wharf."

"So this is your vaunted New England morality."

"Yes sir it is. Speak to our good man William Bartlett he also severely questions our morality."

The colonists laughed while unloading 52 chaldrons of coal, 86 butts and 30 hogsheads of porter, 20 hogsheads of vinegar and 16 hogsheads of sauerkraut, They also led several head of cattle down the gangway, which would be sorely missed by the redcoats garrisoned in Boston.

The redcoats were much plagued with hunger and smallpox. They were entirely dependent on food imported from England, since none of the colonial farmers would sell them produce.

Chapter 15
"Never Trust a Privateer"

Things got to be so bad that on March 17, 1776 General Howe retreated from Boston and withdrew his starving soldiers to Halifax, where the locals would sell them food.

There was general rejoicing throughout the colonies. The good men and privateers of Newburyport had done their part.

CHAPTER 16
The Timber Ship
December 1769

Geoffrey and Michael stood at the stern watching Plum Island disappear in a cloud of swirling snow.

"Do you think you'll ever see it again?"

"Plum Island? Aye, I'll be swimming there again next summer!"

"In this old scow? It looks like nothing but an Archangel pram."

"I'll grant you she doesn't look like much. But I warrant she'll prove sound. Curriers men know how to build ships."

"But this is just a bunch of unmilled logs. They were floating down the Merrimack only a few months ago."

"But you watch. She will lie low in the water and be carried to England in a river of warm blue seawater."

"Nonsense! Here we are risking our lives so our fancy dry goods owner can avoid paying any taxes on cargo. This is just a way to get around crazy King George."

"Never underestimate Mr. Levi and our good Captain, Mr. Rose."

A few days later the good captain saw what he was looking for. A squall line marked the boundary where the dark green waters of the Atlantic gave way abruptly to the pale blue waters of the Gulf Stream.
The temperature rose to 70 degrees and the sailors took off their coats. They would enjoy these balmy conditions for the rest of the trip to England.

The following year Benjamin Franklin would write that if captains positioned their ships in the mighty river of seawater he named the Gulf Stream they could deliver the mail several days sooner to England. But experienced sailors already knew to use the northern route when sailing to England and the southern route when sailing home.

"Ethan have you noticed that there are fish swimming in the shade under our hull?"

"Really! How do you know?"

""You can see them between the logs."

"Cook said he saw a shark yesterday. It was eating the slops he had just thrown overboard."

"I'll bet if we dangled a hook over the side we could catch some fresh fish for supper."

"Aye, I've been thinking too. If we fashioned a net from some sailcloth we could probably get ourselves some turtle soup and I could take some of the shells back to Waterside so the Noyes brothers could make Sarah a turtle shell comb."

"And wasn't it you who said we would never get back to Newburyport on this stinking raft?'

With such bantering the crew made their way across the Atlantic in an astounding 26 days. They tied up at Orchard House Black Wall where the 300-foot ship fashioned from timbers joined together to conform to the shape of a ship received widespread attention. Londoners marveled at the cavity left in the center of the timbers where the crewmembers ate and slept. What would those damn Yankees think of next?

Chapter 17
The Bridge
1806

Benjamin Clifford loved to regale his guests with tales of the old fort that used to sit not far from his small hotel on the northern tip of Plum Island.

"Oh yes, Fort Faith was an active place during the Revolution. Newburyport had some of the best privateers on the East Coast. Got their start building a series of secret underground tunnels so they could avoid paying British taxes. Made much more sense than throwing good tea into the harbor, like those fools in Boston."

"If I were in charge I would have had the Privateers lure one of the British ships into the river then catch them in a crossfire between Fort Faith and the batteries on Salisbury Beach."

"But then they would have been close enough to fire on the town."

"Didn't matter, we had several small ships on the river that could have captured them before they had a chance to turn and run."

"Sounds pretty risky to me."

"Would have worked if the Brits had even the slightest interest in attacking our town. Newbury probably would have cheered. There was no love lost between the two towns in those days. You see Newburyport had grown so fat and rich from smuggling goods they had split off from Newbury in 1764. Newbury has remained a jerkwater little town ever since.

"Oh come now, Clifford!"

"Well the towns did manage to get back together long enough to fund building a lighthouse on the end of the island. Before that people would just burn wood fires or wave torches around on the end of long sticks."

"Sounds like mooncussers to me."

"What were mooncussers?"

"They were brigands who rode down the beach waving torches in the air so they looked like small boats sailing back and forth in the harbor. The sailors would see them and turn into what they thought was the river only to end up shipwrecked on the outer bar. The next morning everyone would be out picking over the free bounty. Only worked on nights where there was no moon, so they called 'em mooncussers."

"Maybe they had such on Cape Cod. Up here we had none who would purposefully cause a ship to run aground. Mind you we had wreckers, but they were professional men. If your ship happened to run aground the wreckers would appear out of nowhere offering to get you off the bar for a fee."

"Hmm for good money I would suppose."

"Well in an emergency you are willing to pay almost anything, aren't you?"

"It must have been a boring existence being a lighthouse keeper."

"Not during an emergency. The keeper had to raise the distress flags and fire a cannon to alert the townspeople that help was needed."

"They also had to rescue and care for any survivors. Erosion kept them busy too. The end of Plum Island grew north every summer but vicious storms often eroded it back in the winter."

"They kept the lighthouse on blocks so it could be moved to adjust for erosion. Captain Hunt wrote that he had to move the lighthouse two times and the bug light four times during his brief tenure out here."

"What on earth is a bug light?"

Chapter 17
THE BRIDGE

"They were small lighthouses that were placed on the beach about a hundred feet east of the main lighthouse. The pilots knew they had to line up the two lights as they entered the river, otherwise they were sure to hit the bar."

"Things got easier when the Merrimack Humane Society built two life savings stations on the island. Every night a man from the Merrimack station would hike 4 miles down the beach looking for shipwrecks and survivors. He would meet the man from the Knobbs Beach station who had been hiking the other way. It was cold and sometimes several feet of snow covered the slippery dunes, making it almost impossible to walk."

"One night one of the men didn't return from his rendezvous so they sent out a search party that found him frozen and permanently blinded by the pelting sleet. From then on Captain Hunt advised his men to hold a wooden shingle in front of their face when they went out in such conditions."

"Why even last year before the bridge was built, the sloop *Blue Bird* got caught in a Northeaster carrying freight from Boston to Newburyport. The winds pushed her into the curve of Plum Island where she ran aground on the offshore bar."

"But the light keeper saw her flares and alerted the townsmen who had rowed over to the island to help. They could see the Captain and crew huddled along the rail as waves kept pounding the *Blue Bird* onto the bar."

"The keeper used a giant megaphone to advise the men to jump into the waters before the sloop broke totally apart. Fortunately all the men could swim and it was October so the water was still warm. One by one they all jumped in and the waves helped push them onto the beach where they were given warm blankets and brought back to the keeper's house where his wife had built up a nice warm fire."

"The next day the beach was littered with $100,000 worth of books, hardware and woolen goods. To this day you can still see wooden paneling from the Blue Bird in the parlors of some of the finer old houses on High Street."

"That was in 1805, the following summer a group of Newburyport businessmen built a bridge to the island and we built this hotel out here on the point."

"Yes it's outrageous they charged us 4 cents to ride across that little drawbridge!"

"Oh you got a deal. They charge farmers 25 cents to bring over a hay wagon and 5 cents for a dozen pigs."

"Preposterous!"

"But we're thinking of digging a canal so we can ferry guests from the river to the hotel."

"So you will be entering the modern age."

"All except for plumbing, you cant have that on an island such as this."

"Why not?"

"Salt water would get into the pipes and corrode them. Nope outhouses and thunder pots work just fine."

Chapter 18
The Pocahontas
December 24, 1839

'Twas the night before Christmas and Mrs. Lunt couldn't sleep. Nine days before, a storm and high course tide had overflowed the Riverside wharfs and flooded the Lunt's house on South Street.

Breakers had washed over the northern tip of Plum Island leaving such a large lake between the lighthouse keeper's dwelling and the lighthouses that he couldn't tend to the twin beacons. The surf flattened 20-foot high sand dunes and surrounded the Plum Island hotel with foaming water. Now the beach was several rods narrower than before.

"What more can possibly happen out there, Cutting?" Lunt snored in response.

Just yesterday, the winds had swung around to the north blowing the wispy fog out of its proper home in the Bay of Fundy and down to cover the coast of Plum Island. There it shrouded the surf, breaking over the bar and dimmed the warning harbor lights.

The drizzle had turned this into a proper tempest of pelting snow and sleet with the sepulchral roar of the Atlantic chilling Lydia's very bones.

Lunt belched, broke wind and rolled on his back. Lydia's elbow brought him to muddled consciousness.

"Cutting, listen to that urgent sound. I can hear it in the surf. It frightens me." Cutting had gone back to sleep.

The next morning the Lunts joined a score of men and women gathered on Plum Island to witness the frightening conflict of the battling elements. Hattie spotted her friend and ran to her side.

"Oh Lydia it is so awful a 300 ton brig struck the bar last night and there is nothing that can be done to save the men who still cling to the wreck."

"Lord a Harry, how dreadful!"

"Those aboard who still have life can see us on the shore, but we are powerless to give them aid."

"God help them. "

"The surf is too strong to launch a boat and it is too distant for the men in the brig to throw a line ashore."

"I see one man clinging to the bowsprit."

"Indeed, he has been there all morning."

"Oh dear Lord, a giant wave just washed him and the bowsprit away. Can't anybody do something?"

"It would just lead to more loss of life," intoned Lunt at her side.

Just then the storm surge lifted the brig off the bar and the waves drove her quickly onto the beach so the helpless onlookers could sweep aboard.

"There is still one man lashed to the mast."

"Cut him down."

"It is too late. He expired in my arms too exhausted from the constant pounding to utter a final word. The waves had beaten over him so fiercely that his clothes had been stripped entirely off his naked body."

"Oh Hattie you do say the most unsettling things!"

The storm continued unabated, driving ice floes across the flats and over the wharfs and piling them up on Water Street.

Chapter 18
The Pocahontas

Even at low tide the ocean flooded the blocks under the twin lighthouses so the keeper feared the beacons would topple into the seething foam.

When the tide went out, the shore party found the lifeless bodies of nine sailors lying white and shining in the moist sand. Nearby, was the overturned boat they had tried to use to escape the brig the night before.

In all the rescuers recovered the bodies of seven crewmembers and two officers. They were laid out in the broad aisle of the Old South Church and 2,500 mourners filed past pausing to reflect on the features of the young Captain Cook and his cousin who had grown up in town and the seven unknown men whose families would never know their fate.

Bells tolled throughout Newburyport and flags sagged listlessly at half mast as the pall bearers carried the young men up to the hillside burying ground.

Every year Lydia and Hattie would trudge up that sunny slope to pause and reflect on the monument built to honor the young men lost on that terrible night, and remember the young men they had loved and lost in their long gone happy youth.

Chapter 19
Parting the Waters | Moses Pettingell
1850

Life proceeded on Plum Island through the early 1800s, a few harsh winters, a few mild winters and a few highly unusual high tides. Other than that, the natural cycles of wind, weather, storms and seasons continued in their magisterial pace and Fort Faith continued to protect Newburyport's growing fleet of smugglers and privateers in the War of 1812.

But strange things were happening on Joppa Flats. The Pettingell brothers were unusually close. Their father had willed that Moses care for his handicapped brother Eleazer. So, after Moses returned from the Revolutionary War , the two lived together in their family home making a decent living fishing in the summer, hunting seagulls and shorebirds in the fall and making shoes in their kitchen all winter.

It was the shoes that brought in the hard cash so they could afford to pay for a housekeeper Sally Beckett. The life of the three was shrouded in mystery until Eleazer married Sally in 1795. She was already quite pregnant with a baby that the three named Moses, which occasioned considerable speculation among their neighbors about the lad's actual paternity.

Be that as it may, the three determined to be perfect parents. They doted on the young boy providing for his well-being and an exemplary Presbyterian upbringing.

The first thing they did was build a larger home with two dining rooms, two living rooms, and several common rooms to house their farm workers. But the main feature of the grand house was that it provided a spectacular view of the north end of Plum Island. This would prove to be propitious.

Much of the young boy's life centered on attending the Old South Presbyterian Church and it seemed to work. By all accounts he grew to be a kind and generous person with such a good heart that people started to call him Squire Moses, though his son Warren found his father to be a tad too sanctimonious for his wilder tastes.

Part of Moses' farm holdings consisted of salt hay meadows on Plum Island that gave a salty tang to his cows' milk. Then, in 1829 Squire Moses bought the entire Northern end of Plum Island and arranged to have it and Joppa Flats become part of Newburyport.

His son Warren loved to escape the drudgery of his father's mainland farm by leading a cook and crew of farmhands to Plum Island every summer. They would set up tents and spend their days scything hay and their nights playing cards and drinking great quantities of hard cider. In the fall they would load the dried hay onto Warren's gundalow and he would sell it up and down the Merrimack River.

In the off-season Warren would cut lumber and deliver the "stickes" to the mast works. The stickes were the reason that Newburyport's main streets that run up the hill from the river are so straight and wide. They had to be able to accommodate the oxen hauling the tall trees to the mast works on Federal Street or up the hill to the old Boston and Maine freight yard.

But the way Moses made his real money was by mining for sand on Plum Island.

~ The Lawsuit, 1883 ~

"God damn that Boynton Moody! Who the hell does he think he is?"

"Moses wouldn't like to hear you cuss like that!"

"Yes but it was grandfather who bought Plum Island."

Chapter 19
Parting the Waters | Moses Pettingell

"Remember him standing in the house on Bromfield Street peering through that 20 inch telescope as his men on Plum Island pushed wheelbarrow after wheelbarrow of sand up a narrow plank then dumped it into the sand schooners?"

"What did they do with all the sand?"

"They sold it to builders in Boston who were making all those elegant houses in the Back Bay."

"Too bad so many of them are falling down."

"Shush, don't say that!"

"Why not?"

"The reason they are crumbling is that the Plum Island sand contained too much salt which made their mortar unstable."

"Did grandfather know that?"

"No, nobody knew it at the time."

"Then what happened?"

"Well remember those storms back in 1839? They cut right through Salisbury Beach so there were two channels with an island of sand in between them."

"I think I read that the early settlers found about the same thing back in the 1600s."

"Yes some think that every few hundred years waves push enough sand up the beach so it seals off the mouth so the river jumps north."

"How does it do that?"

"Well first some storms wash through the beach so two channels are formed but a two channel system isn't very stable. Then over a few years, sand flowing up from the south starts to fill in the southernmost channel and the northern channel starts to dominate and become the new mouth of the river."

"So you mean that what we call 'The Basin' used to be the old mouth of the river?"

"Exactly, but what gets Moody's goat is that when the new channel formed, about half a mile of Salisbury Beach was added to grandfather's end of Plum Island. But Boynton claims he bought the beach from some Salisbury proprietors."

"Does he have a case against us?

"Not a bit of it. According to good old colonial law if a beach attaches to your property, it is your good fortune but if it washes away, it is your loss. You see you can never really own a beach you simple own an area where the sand is resting for awhile until it continues its journey to somebody else's area. So you own a process, not a piece of land."

"So haven't you become quite the lawyer! But what does British law say about the land beyond low tide?"

"Oh that's the 'Queen's Bottom,' anyone can scratch clams off the 'Queen's Bottom!'"

"Hah, hah, hah. Good one, good one!"

'That's what a year of law school and no degree will get you, a smattering of legal trivia, excellent fodder for cocktail parties, not enough to make an honest living!"

"Speaking of cocktails, shall we join the ladies and hoist one to good old Grandfather Moses?"

(Exit laughing.)

Chapter 19
Parting the Waters | Moses Pettingell

Chapter 20
To Move a Lighthouse
1869

Lighthouses have always been problematic on Plum Island. A tornado toppled the two twin houses in 1808. But when the mouth of the Merrimack moved north after the 1839 storms, the remaining wooden lighthouse and its bug light had to be moved as well:

"Hold the horses steady while we ease 'er onto the sledge."

"They're ready to pull this morning, I'll tell you."

"Fine, now John, can you hold your barge steady against the sand while Sebastian pulls the sledge aboard?"

"I can!"

"Good, now heave, heave!"

When the lighthouse was safely on the barge, the men pushed, poled and sailed it to the other side of the basin.

"What are the coordinates for the new foundation?"

"42 degrees, 48 minutes, 59 seconds North, 70 degrees, 49 minutes, 6 seconds West "

"Fine men, leave 'er here boys. We'll fetch the bug light first thing in the morning."

Henry Hunt would have to move the lighthouse two more times and the bug light four more times to keep up with the fast-changing river mouth.

But Henry's life became a little easier when the Federal government established the Life Savings Service in 1871. They built a small station on the High Sandy drumlin about a mile south of the center of the island. But its location turned out to be too far from the river mouth where the ships foundered on the harbor bar as they attempted to enter the river.

The station was moved back to the river mouth in 1881 where it worked in tandem with the Plum Island Life Savings station that was built near the southern end of the island at Knobbs Beach.

Life was particularly harsh in the Knobbs Beach station during the winter months. Every night one of the crew members from Knobbs Beach would have to walk four miles north through soft sand and snow to fetch a metal tag that proved he had walked the entire distance looking for stranded vessels.

A crewmember from the Merrimack Station would trek four miles south to do the same thing. While another Knobbs Beach crewmember would have to retrieve a tag from the Bar Head key post at the southern end of the island.

The crews participated in over a hundred rescues because sailing ships would be routinely blown downwind to wreck on the island, before the advent of steam engines made it easier for ships to stay offshore.

But travel was safer on the river. By 1876 as many as ten steamers carried passengers downriver from the mill towns of Haverhill and Lawrence so they could set up tents and enjoy a day on Plum Island. Mrs. Vale Smith remembered seeing the scene.

"The sandy beach was dotted with tents and around them were groups of young men and maidens, old men and children. The complacent pastor, the grave deacon, all enjoying together a day of unrestrained mirth and healthful recreation."

Chapter 20
To Move a Lighthouse

Other tourists took horse drawn buses from Newburyport's Market Square over the causeway to stay at the Plum Island Hotel and the Bay View that had been built in Salisbury and later moved across the Merrimack by barge.

Homeowners and visitors from Ipswich took the steamer *Carlotta* to get to Sandy Point and Grape Island. So, by 19th Century, Plum Island was a thriving beach community with dance halls and pavilions connected by a railway at one end and quiet farms and cottages at the other end.

Chapter 21
The Mystery of the Missing Matriarch
1879

In 1879 three gentlemen decided to leave their party of beach goers and wander down to the Knobbs Beach Life Savings Station. But on the way one of them saw something strange jutting out of a sand dune.

"What the devil is that?"

"Why it looks like some kind of tusk. Maybe from a whale or walrus."

"Don't be absurd whales don't have tusks and walrus no longer get down to these parts!"

"Lets get some shovels and dig it out."

A few hours later they had unearthed a skull between two and three feet wide, seven feet worth of backbone and a massive leg.

"My word it looks like some form of elephant."

"Indeed that leg bone must have had enormous solidity when it was attached to the living animal."

"From the condition of the bones they must have been buried for ages."

"Yes they are already starting to crumble."

"What an extraordinary find. We should contact Dr. Agassiz at Harvard. It might be one of those Ice Age creatures he keeps talking about."

"Killed by the great deluge, I should think."

Indeed the three gentlemen had discovered the remains of the matriarch of the herd of woolly mammoth whose body had come to rest beside one of Plum Island's ancient drumlins.

Dr. Agassiz was thrilled to hear of the find.

"Precisely what I thought. A creature of the Ice Age no doubt."

"But how could such an ancient animal end up on such a young beach?"

"Now I don't want you to go theorizing about evolution and such. That Darwin has been getting far too much attention for his cockamamie ideas."

"But it would make a perfect specimen for our new museum collection."

"Indeed we could display it beside the elephants of Africa and India and show how an Ice Age creature could never be related to a warm climate animal like an elephant."

It would take 150 years for scientists to finally figure out how you could find a 13,000 year old woolly mammoth on a 4,000 year old barrier beach, to say nothing of how you could have a 13,000 year old fossil if the world was only 6,000 years old, as so many of Darwin's opponents claimed.

The other great question was whatever happened to the body in question—the missing evidence for evolution and the Ice Age.

Harvard's Museum of Comparative Zoology, established by Agassiz to refute evolution, had no record of the find. They have also changed their name to Harvard's Museum of Natural History, so as not to remind anyone that one of the college's most celebrated professors had been on the wrong side of the evolution debate. Though many think it is the sign of a great university to have at least one faculty member on the wrong side of every major intellectual discovery.

So perhaps the remains of the matriarch were quietly disposed of during the night, or they molder on in the attic of one of the old High Street Captain's houses in Newburyport.

Chapter 21
The Mystery of the Missing Matriarch

But that is not the end of the story. The matriarch would rise again to become the center of another great controversy, as we shall see in 2013.

Chapter 22
Island Resort
1880s

Little by little people started to build summer cottages on Plum Island. In 1880 Michael Simpson built an imposing cottage just south of Plum Island Center that still survives as Bennett Hill.

The Pettingells also started selling lots so more people could build small camps and grander cottages. One of these was given the name Rinko-Tei by a guest who had been ambassador to Japan. It means the place between the waters, which fit its location between the Basin and the mouth of the Merrimack River.

The Plum Island Hotel initiated coach service from the train station in Newburyport to the large Victorian hotel during the summer months. But the real change came about on May 9th 1887.

Maggie: "Father, is it time to go see the new horse cars?"

Tom Duncan: "Yes Maggie, go get your sister. We don't want to miss them."

Laura Duncan: "I've never seen so many people in the Center."

"Here they come! Here they come!"

Four beautiful chargers pulled the horsecars down the iron rails from the south end of Newburyport to Plum Island Center.

"Oh Father, now we can stay out here all summer and you can take the horsecar to work!"

"I suppose you can as long as I don't go broke spending 5 cents a day for a 20 minute ride."

The horsecars proved to be a grand success, sometimes running as many as 14 open horsecars a day jammed packed with beach goers and sightseers.

The horsecars ran until they were replaced by open electric cars and a freight trolley that delivered coal, ice and food right up to the broad veranda of the Plum Island Hotel. The hotel now boasted 48 rooms and a bustling dining room that serve up to a hundred people a night under the management of D.H. Fowle.

After dinner hotel guests could hear a band concert or go dancing at the nearby Pavilion with its 10 foot wide wraparound piazza. Some might want to catch a film at the movie theater near Plum Island Center or dance in the new ballroom. As many as a thousand paid admissions were not unusual at the ballroom on a busy Saturday night and during the week the dance floor was converted to a roller skating rink.

But life proceeded at a more leisurely pace at the isolated southern end of the island. You could only get to the small hotels on Grape Island and Ipswich Bluffs aboard your own boats or on the *Carlotta* steam ferry.

The Lufkinds ran the hotel on Grape Island and charged 25 cents for use of their tents. Guests lived in the tents, ate in the Inn's small dining room where bowlfuls of steaming chowder were served along with the farms own milk, corn and eggs. Beachside clambakes were put on every weekend.

For special occasions, Ipswich Bluffs Inn could accommodate hundreds of people. On Rowley's 137th Anniversary the steamer *May Queen* brought over 1500 celebrants over along with the Boxford Brass band. The extras were towed behind the steamer in scows. But a delicious time was had by all, and nobody needed to worry about driving home three sheets to windward.

Chapter 23
The Jetties
1881-1900

After the Merrimack River jumped north, the Army Corps of Engineers was brought in to decide what to do. First they tried building a dike across the Basin so the river couldn't decide to course back down her old channel again. The dike created a safe place to swim and the Basin became a favorite place to build summer cottages. But the dike gradually fell into disrepair, although its remains can still be seen at low tide.

The more ambitious project was to build two jetties to both deepen the mouth of the river and jettison away the harbor bar. Jetties had proved their worth after the Civil War when a huge muddy sand bar had made it impossible for ocean-going vessels to enter the Mississippi River.

The Army Corps was considered to have the best professional engineers in the country and it was led by General Humphreys who was a considered to be the world's expert on the Mississippi River.

He was also a hero of the Civil War who was known for his ruthlessness. During a single charge he had lost 20% of his men in 15 minutes, and wrote that the experience made him feel like a 16-year-old girl at her first party ball.

He was equally determined to crush James Eads, a self-taught engineer who threatened to topple the comfortable monopoly on engineering that the Army Corps had enjoyed since its incorporation under George Washington.

Eads gained most of his knowledge of the Mississippi River by walking along its bottom under a heavy salvager's diving helmet. He knew what it felt like to grope through billowing clouds of thick brown silt and be pushed and pulled by underwater currents, never knowing when they might suck you into a newly scoured death hole.

He had used his river knowledge and engineering skills learned late at night by candlelight, to design the St Louis Bridge, the first bridge made entirely of iron. The bridge had made him famous, so when Congress called he was ready,

General Humphreys had tried the traditional method of dealing with a harbor bar by dredging it, but his dredges had become stuck in the sticky, thick Mississippi River mud. He was furious when Eads told Congress he was so sure that jetties would work he would build them at his own expense and would only be paid if they proved successful.

The dispute came to a head on May 12, 1876. General Humphreys had leaked information to the press that the Corps had made soundings of Eads's channel and it was only twelve feet deep. The news had caused stock in Eads' company to collapse, making it impossible for him to borrow more money.

Eads knew his only chance for redemption lay in the oceangoing steamer that lay just offshore. It was the *Hudson* a three-hundred-foot vessel that drew 14 feet, seven inches.

The *Hudson* was under the command Captain E.V. Gagner, Ead's old friend from their early days on the Missouri River. Gagner invited Eades aboard, along with several journalists Eads had invited to chronicle his gamble. Everyone knew the stakes were as high as at any riverboat poker table in New Orleans.

Gagner also knew how dangerous the situation was. The tide was falling fast and his local pilot recommended that he not attempt to cross the bar. But Gagner did not hesitate.

"Head for the jetties."

On shore three hundred men ceased their labors to watch. They too knew the stakes as the ship started to pick up steam.

"Shall we run in slow?"

Chapter 23
The Jetties

"No sir. Full speed ahead."

Gagner knew that the increased speed would lift the *Hudson's* bow a few inches above the surface and push her stern a few inches below it. If Eads were correct she would just skim lightly over the mud. If the Corps were correct the *Hudson* would tear out her hull and sink to the bottom.

The *Hudson* gained more speed and a huge white wake billowed out ahead of her bows then separated into a long "V" that sloshed to the edges of the willow-sided jetties. As one of the journalists wrote, *"As long as she carried that white bone in her teeth. The great wave that her brave bows pushed ahead of her as she sped forward – we knew that she had found more than the General Humphrey's 12 feet."*

Then she was through. Captain Gagner blew a powerful blast on the *Hudson's* steam whistle and three hundred men erupted into cheers that reverberated up and down the delta as the great ship made her way majestically upstream.

CHAPTER 24
The Wreck
1886

Breeches buoy, Courtesy Custom House Maritime Museum, Newburyport MA

In early December a Northeaster stalled off Plum Island and hurricane force winds lashed the shore from Salisbury Beach to Gloucester. Heavy wet snow and piercing sheets of sleet cut visibility to a scant few feet.

Frank Stevens was repairing some gear in the Knobbs Beach Life Savings Station when he saw the naked masts of a schooner loom through the clouds.

"Lord a Harry! Watson we've got to get out thereah. That vessel will be in 10,000 pieces before mornin'. It may be our only means of savin' life."

"But there is eight feet of snow in front of the boat house dooah, sir."

"Have the boys dig it out and hitch Mabel up to the boat cart."

"That poor old nag wont make it through the storm."

"You know the Life Savings Service motto don't you Nathaniel?"

"Yes sir, 'You have to go out but you don't have to return.'"

"Well that goes for horses too. Now have the boys shovel a path in front of the boat cart and have them hold a shingle in front of their faces when they are not shovelin'."

For the next few hours the men tried to keep Mabel on the faint track that wound through the windswept dunes.

"Mabel is down! Push her back up. The poor thing is almost dead."

"We all are Frank. We are mostly pushing the cart ourselves."

At 2:25 the crew spotted the stricken vessel. She was two miles north of the station and 600 yards offshore.

"Her masts are still standing and her head's to the wind, so her anchor must be holding."

"But her hull is already half submerged. Look at those waves sweeping down her deck. Nobody could survive in those waves."

"Should we try to launch the boat?"

"No sense in that, waves are too high. Break out the Lyle gun and see if you can fire a line into her rigging. If anyone is aboard he can haul it aboard so we can set up the pulleys and winch the crew ashore in the breeches buoy."

But the Lyle gun was useless. The ship was too far offshore and there was no sign of anyone lashed to the rigging. No bodies were ever found from the disaster.

At dawn, Stevens sent a search party into the dunes to see if they could find the patrolman who should have returned from the Bar Head key station for the midnight watch.

Chapter 24
The Wreck

They found the poor soul wandering exhausted and disoriented in the dunes. He had been permanently blinded by the blowing sand and snow in a storm that had taken dozens of lives on both Plum Island and Cape Cod.

Chapter 25
The Gunners
1891

It was a cold drizzly day in early April. Dark clouds hung low in the leaden sky and three men huddled in Bob Wilkinson's boat as it drifted off Emerson Rocks.

Bob tweaked a line tied to several live brant swimming ahead of the boat. Some of these live decoys had been wounded in last year's hunt and Bob had cared for them all winter long. Now they were fat, happy and ready for the spring shooting season.

Bob tweaked the line again and the live decoys gave their inviting cc-cronk, cc-cronk, cc-cronk feeding call. The wild birds took notice.

Now came the hard part. Bob could see his sports were getting excited as hundreds of large birds swam toward their camouflaged boat.

"Stay down! Stay down! Wait until the first ones swim right up to the boat."

"They will be right in our laps."

"That's the idea. Now wait, wait."

Bob finally pulled sharply on the line and the live decoys scurried to the side.

"Why those dirty little traitors to their breed. They do exactly as they are told."

"Now, Now! Shoot! Shoot!"

Bob bagged 40 of the small geese on his first shot. John Phillips and the "sports" shot a satisfying 30 birds each. They had killed 375 brant in the last nine days, a pretty good draw for the month long hunting season.

Back in camp, Phillips recorded the temperature, weather conditions and how many old and young birds they had killed. It was easy to tell the young brant because they still had white edging on their coverts and secondary feathers.

Professor Phillips taught ornithology at Harvard, lived in Ipswich, and was the world's expert on black ducks. At night he and Bob would discuss what the numbers meant.

"Well I'll be damned. Last year we only shot a few dozen birds and we thought they would never recover, but this year we shot hundreds again."

"It must be those bloody confederates who overshoot them in the Carolinas."

"We can't blame everything on the Southerners. No, I think the problem lies in their breeding grounds above the Arctic Circle."

"Have you noticed that we always seem to shoot more young birds during the plentiful years?'"

"Good God, you know I think you've got it."

"Got what?"

"The answer! Some years the Arctic must have more storms in the summer than others and those storms must kill off most of young chicks."

John Phillips had it half right. The cause of the rise and fall of brant did lie in their Northern breeding grounds. But the problem was lemmings, not storms.

The lascivious little rodents would reproduce prolifically every summer until they had outstripped the tundra of moss. Then the entire population would be griped with a migratory mania.

Chapter 25
THE GUNNERS

During their migrations they would often swim across small lakes. But many of them would drown in mid-lake, leading to the legend that lemmings commit suicide by leaping off cliffs and swimming into the Arctic Ocean until they drowned of exhaustion . Anyone who grew up in the Fifties know this to be true, because we saw it in the Disney's film *White Wilderness.*

However it turns out that the lemmings weren't cooperating the summer they were filming so the director had an assistant sitting off camera hurling lemmings into the Arctic Ocean. We always suspected there was something off about Uncle Walt.

So if lemmings didn't leap off cliffs into the ocean, what did cause their fluctuations to occur in regular four-year cycles? First the lemmings would reproduce rapidly until their population peaked also supporting more predators like Arctic fox, snowy owls and weasel-like stoat.

But after about four years the lemmings would have consumed all the moss triggering their migratory zeal and population crashes. Their absence would reverberate through the food chain causing Arctic foxes to move off the tundra to find other food and snowy owls to fly south to feed on voles whose populations were more stable because they fed on fast -growing grasses.

So during the years when there were few predators, the population of brant would surge, because unlike most birds they can lay up to 30 eggs in a clutch. Apparently they also picked up the habit of nesting closer to snowy owls whose talons would keep predators like fox and stoat at bay.

But maybe John Phillips wasn't so far off after all. There is one more theory that puts the fluctuations of lemmings back on the weather. This theory says that when the snow is light and fluffy, lemmings can build tunnels that protect themselves from predators. But when the snow is wet the advantage switches back to predators, because the heavy snow collapses the lemming tunnels.

Some scientists think that global warming is to blame for the wet snow that has caused lemmings to go into a more permanent decline. This has caused snowy owls to irrupt south more often, to the delight of today's Plum Island birders.

However, by the time Phillips wrote up his findings in the 1930's however, birders were far more concerned about regulations. Spring shooting of ducks and brant had been banned in Massachusetts in 1909, and in 1912 market hunting was made illegal along with the use of live decoys.

As early as 1879 a well-known egg collector and ornithologist had led birding trips to Plum Island aboard the *Carlotta* out of Ipswich. His walks became highly popular, attracting birders from all over New England. Then as now, there was a spirited rivalry between the different groups of bird clubs.

But through the walks, all the clubs and birders became aware that Plum Island was a crucial link in the Atlantic flyway, and one of the best places on the East Coast to see up to 300 different species of migrating birds.

Edward Forbush, the state ornithologist for the Commonwealth of Massachusetts wrote and a frequent participant on these walks wrote, "Secure Plum Island and make it a bird sanctuary, for in my opinion, it is the most important region on our coast."

So it was no accident that when the heiress Annie Brown bequeathed a large sum to the Federation of Bird Clubs, they decided to use the money to purchase the southern end of Plum Island and establish the Annie Brown Wildlife Sanctuary.

And whom did they hire to protect birds from poachers? The former market hunters Clifford Brocklebank and Charles Safford who used to love to gallop up on horseback and read island visitors the riot act about poaching and gunning.

Chapter 25
THE GUNNERS

The only hunter they never caught was Bob Wilkinson. After suffering a severe stroke, his friends still took him out so he could sit on the running board of the car and watch them hunt pheasants. But one day they heard two shots and ran back fearing the worst. Bob sat in his chair smiling and because he couldn't speak pointed to the left and the right. His friends released their hunting dogs and each came back with a nice fat pheasant in his mouth.

Chapter 26
Rum Runners
1924

The backside of Plum Island is a world apart, a variegated landscape of creeks, marsh grass, and towering phragmites reeds. It stretches from Woodbridge Island south to Ipswich Bluffs. Indigenous species of toads, rabbits, coyotes and deer inhabit tiny thickets of vegetation that thrive in the swales of the ever-changing dunes and former drumlins.

By day, hollows in the dunes act like natural solar collectors. The sun reflects off their sides, concentrating its rays on the floor of the depression. In the summer their sands reach foot-scorching temperatures that rival those in many deserts and mold a singular assortment of plants and animals.

It is also the habitat for numerous endangered species. Piping Plovers and Least Terns nest in new washover areas and tiger beetles scuttle over the dunes.

Equally endangered were the humans who used to call this land of marsh and creeks home. Elizabeth lived at the end of a little traveled sandy road that wound through the extensive marsh. She kept her boat in a deep creek that flowed past her lone cottage toward the Parker River and Ipswich Bay beyond.

In early October she was hurrying through the sandy twin tracks that cut through the bogs of Hell Cat Swamp. She loved this spot. Ever since her husband died she had made her living raking cranberries, collecting beach plums and digging clams.

She wanted to get home before dark to cook up some of the sand dabs she had caught off Emerson rocks. She paused to listen to a copse of phragmites that rustled in the wind above her head. It was a scratchy, ethereal sound that matched her restless mood. She imagined the terns felt the same way before they started their long migrations south.

Elizabeth built up her stove and her camp was soon filled with the smell of the fresh caught flounder and cranberry muffins she had baked the day before. The dishes could wait for tomorrow. She turned down her kerosene lamp and crawled into her wooden bunk bed.

But later that night she heard the sound of a long, sleek, black boat putter up to her dock. She knew a mother ship full of Canadian whiskey had been lurking offshore for the past few days. Lights flickered and a truck coasted to a halt outside her camp. Then she heard voices inside her kitchen. She reached under her bunk and felt for her husband's old shotgun.

"Who's down there? What's going on?"

"Don't worry lady. You have nothing to fear. You just stay upstairs and we'll make it right for you."

She couldn't make out any more words but figured there had to be at least three or four men, a boat and a truck parked beside her camp. She huddled back down under the blankets and waited until the house was quiet for several more hours before venturing downstairs. There, sitting on the edge of her kitchen was a crisp new hundred-dollar bill.

"Now isn't that nice of those kind gentlemen. Next time they come I think I will make them some piping hot cranberry muffins."

It is said the mutually beneficial business arrangement continued for many profitable years.

Chapter 27
Rachel Carson
1946

Parker River Wildlife Refuge by Rachel Carson, Courtesy National Wildlife Service

In 1946 two young women rode on the tailgate of the new Parker River Wildlife Refuge truck as it bumped and swayed down the beach toward High Sandy. The ranger gestured to the south.

"If you climb to the top of that old Army Observation tower you can get a pretty good panoramic view of the island."

Rachel Carson scribbled in her notebook, "To the east is the vast immensity of the Atlantic, nothing between us and Spain," then turned to her friend and illustrator Kay Howe.

"If I were a bird I wouldn't care about the Atlantic. I'd be interested in the five feeding zones you can see as you look south, the ocean beach, the dunes behind the beach, the thicket of small trees and shrubs that runs down the center of the island, the salt meadows, and the tidal flats in the salt marsh creeks."

"It looks like a bomb exploded in those blowout areas. Nothing growing on their bare walls. But I'll bet spring migrants can feed on the bayberry and poison ivy seeds."

"See if you can get a picture. It could make a nice illustration."

"That would be dramatic."

"The small trees and thickets of vegetation that form a midrib to the island support deer and pheasant. But the whole reason the refuge is here are the black ducks that feed on the marsh."

Rachel turned to the ranger, "How are the ducks doing?"

"Still staggering from the 'duck depression' of the '30's. Last year was almost a complete failure."

"The refuge should be a big help."

"Yes it is the first coastal refuge where the birds flying down from the north can stop to rest and feed."

"As we drove out of town toward the island we saw hundreds of ducks-bobbing like boats in a harbor."

Chapter 27
Rachel Carson

"Tomorrow morning there may be five thousand, the next week as many more. They will spread out in the marshy areas and start gorging themselves on the seeds of the salt marsh plants."

"In the days before the refuge the first roar of guns on the opening day of the hunting season would send the ducks flying out to sea. There they were safe from the barrage of lead but the open ocean provided them nothing to feed on."

"Now the refuge concentrates the ducks in the marsh. They know they can find both safety and food in there."

"Are the hunters upset to lose so large an area for shooting?"

"Quite the contrary. They are already noticing that the hunting has improved on the mainland side of the marsh and the season is longer because the ducks are not being driven out of the vicinity. So, by a seeming paradox, the refuge has improved hunting even while it has helped conserve the ducks."

"What a wonderful thing to write about."

"It is really the lynchpin to the Atlantic flyway."

"What's that?" asked Kay of the ranger.

"There are four major flyways that birds who breed in the north use to migrate south. The birds have a hereditary attachment to one particular flyway and as a rule never transfer from one to the other."

"And the striking feature about the Atlantic flyway is that it looks like a distorted funnel with a narrow stem. All the birds that breed in the North have to fit in that narrow stem of coastal marshes during the winter. So the Fish and Wildlife service has established this string of refuges from New England to Georgia so the birds have a place to rest and refuel every few hundred miles on their journey, and Parker River is the northern entranceway to the coastal flyway."

By this time a small group of eager birders had formed around the two intensely private women. Rachel Carson was already well known from her first book and radio programs about conserving nature. As they bounced back home down the new road that had displaced the heron rookery in Hell Cat Swamp, Kay teased Rachel about needing to buy a disguise.

Chapter 28
The Sea Haven Polio Camp
1947

The Sea Haven Polio Camp

Polio has been around at least as long as Plum Island has been in existence. A stele from Egypt's 18th dynasty depicts a man with a walking stick and atrophied leg swigging down his polio medication, and everybody remembers the PBS tyrant I Claudius limping across their TV screens.

However, the virus that causes poliomyelitis only existed as a fairly rare endemic pathogen until the 1900's when it started to proliferate into widespread clusters in Europe and America. The outbreaks usually started first in cities during the summers, so that families who could afford it flocked to mountain and seaside resorts like Plum Island.

But the most serious epidemics occurred during the 1940's and 1950's. During those years the scourge killed over half a million people and crippled many more. Even America's strapping young president, Franklin Delano Roosevelt was not spared. He died from complications of polio while swimming in the warm waters of a therapeutic pool in Hot Springs, Georgia.

One of the North Shore kids who contracted the disease was Daniel Harrington of Haverhill, Massachusetts. He recovered and went on to become the much beloved head of the Haverhill school system but he always remembered the feeling of being ostracized as a child because of his disabilities. In later life he decided to establish a polio camp, free for kids who were excluded from going to regular summer camps.

So, in 1946 he obtained a ten-year lease on the Knobbs Beach Coast Guard Station and started building a large blue swimming pool overlooking the ocean. The pool became the heart of the Sea Haven experience. Kids from all over the North Shore lived in spartan camps tucked into the dunes and spent most of their days taking therapeutic soaks and playing in the saltwater pool.

The pool was the one place where kids could enjoy a semblance of the mobility that other children took for granted. They would spend long hot summer days forming friendships with other children similarly afflicted with the crippling disease.

In the evenings the campers would gather around the piano in their wheelchairs and belt out their favorite songs as Doc Harrington accompanied them on the camp's well worn piano. One former camper remembers her summers at the camp as among the happiest of her childhood and a former counselor credits the Sea Haven pool as being the place where she first fell in love with the man she ended up marrying.

Chapter 28
THE SEA HAVEN POLIO CAMP

Things were not so copacetic in the world at large, however. When Jonas Salk introduced the Salk vaccine in 1955, there had been 39,000 cases of polio in the United States. By 1956 it had plummeted to 15,000 cases, then by 1961 there were just over a thousand. But the vaccine's great success did not stop the American Chiropractic Institute from mounting a no-holds barred fight against any kind of immunization.

The Institute believed that all diseases were caused by subluxations of the spine and that "chiropractic adjustments should be given to the entire spine the first three days of acute polio infection," and they reported complete recovery in 30% of their cases. But they would not allow scientists to verify their results.

They also concocted a wide-ranging battle plan to argue against immunization. It included 6 major tactics:

One—Spread doubt about the science behind immunization. For instance, they argued that diseases have natural cycles so the apparent success of immunization was just because polio had died out on its own.

Two—Question the motives and integrity of immunization specialists. To do this they argued that there was a nefarious conspiracy between scientists and pharmaceutical companies to make a huge amount of money eradicating polio. If so, who wouldn't be for it?

Three—Magnify disagreements about technical issues like the number of doses to give and whether booster shots were necessary, into a broad-based repudiation of immunization itself.

Four—Exaggerate the potential harm from vaccines, like today's anti-vaxxers who claim that vaccines cause autism, a bogus theory that has been proved wrong by numerous scientists in multiple studies.

Five — Appeal to personal freedom. To do this they argued that compulsory vaccination was another conspiracy, this one against Americans basic freedom of choice. The Supreme Court rejected this argument on the grounds that individuals' beliefs are fine as long as they don't subordinate the safety of the entire community.

Six – When the results started coming in showing that the vaccine had worked, the early chiropractors used their trump card, claiming that Chiropractic methods were a matter of faith and thus not open to scientific inquiry.

The case has particular resonance because many of today's climate deniers have stuck to the same game plan.

One—They blame global warming on natural cycles.

Two – They argue that scientists and environmentalists are in cahoots to get funding for more research.

Three – They magnify reasonable disagreements among scientists about how much and how quickly the sea will rise, in order to dispute that the climate is changing at all.

Four – They argue that slowing global warming will hurt the economy and cost people their jobs. In fact the opposite is true.

Five – They argue that homeowners have the right to protect their private property even when it subordinates the safety of the public trying to swim on the same public beach.

Six – They argue that things like building seawalls and repairing jetties has stopped erosion on places like Plum Island. When in fact houses are teetering on the edge and inconvenient things like woolly mammoths keep popping out of the dunes.

But I guess the climate deniers have it right. Storms are not getting more frequent. The Gulf Stream is not several degrees warmer than it was only twenty years ago, and it is not accelerating storms bulging out of the polar vortex like an old man's giant hernia. If the climate deniers haven't fooled the public, at least they have fooled themselves.

CHAPTER 29
The Groins
1962-1964

Groins

If you look at a series of aerial photographs of Plum Island you can see that like living creatures, barrier beaches need the ability to bend, move and pulsate in order to recover from coastal storms. But humans like to build rigid structures that interfere with such movement.

This was not a problem in the early days when the Plum Island Hotel and the Merrimack River lighthouse were the only structures on the island. But even then the lighthouse had to be moved several times because of ongoing erosion.

It was still not a problem in the 1880's when the first cottages started to be built, because they usually sat on large expansive lots set back from the ocean. When the ocean got too close, the owners simply hitched their camps to a team of horses and moved them to the back of their lots. But eventually things started to get as crowded as a city and there were no longer any beach or dunes between the houses and the ocean.

Then, on September 13, 1950, it finally happened. The Newburyport Daily News reported that, "A devastating high tide backed by gale force winds smashed and ripped the ocean front cottages between 18th and 30th Streets." Two months later hurricane force winds struck again and ocean waves broke through the dunes and flooded down Northern Boulevard, threatening to cut the island in two.

Waves tore a three-story tower off "Ye Town House" cottage during the first storm and toppled the cottage itself during the second storm. Several other substantial homes and cottages were lost as well. Serious consideration was even given to the idea of gradually turning over eroding properties to the Parker River Federal Wildlife Refuge, which was adjacent to many of the houses. Such a move would have prevented many future tragedies.

But town fathers and local construction magnates, who were often one and the same, decided that Plum Island was simply too valuable for tax revenues to simply give away. By 2013 Newbury would be getting 40% of its tax revenue from houses on Plum Island. It was also particularly attractive because the islanders tended to be beyond their child rearing years, so the town didn't have to pay for their children's schooling.

So instead, local officials convinced Congress to appropriate funds so the Army Corps of Engineers could pump sand out of the basin and use it to shore up Island Center. But the Center breached again in 1956, so the state decided to build a series of groins perpendicular to the beach to protect the oceanfront homes.

Chapter 29
The Groins

But the problem with groins is that they interfere with the longshore currents continually carrying sand down the beach. So, while they provided protection to houses upstream of the groins they increased the rate of erosion in front of downstream homes. Plus, waves from northeasterly storms would ricochet off the groins and scour huge scallop shaped gauges out of the beach just downstream of the groins. In fact almost all of the houses lost after the sixties were situated immediately downstream of the groins.

At the same time, nothing had been done about the real problem, that the reservoir of sand built up over thousands of years in the center of the island was gradually petering out so there was less sand available to deal with the effects of sea level rise. Officials had made their first big mistake.

Their second mistake occurred when Mayor Mathews of Newburyport decided that erosion was not getting worse because of anything exotic like sea level rise or Paleolithic sand, but because the jetties built in the early 1900's had fallen into disrepair. His solution was to invite the Army Corps of Engineers back to make repairs.

But as soon as the South Jetty was repaired, it started to act like a 50-foot high dam holding back a wedge of sand close to a tenth of a mile long. Formerly all that sand would have been able to flow through the leaky jetty to build up new North Point, but not anymore. Now scores of houses on Northern Reservation Terrace were in danger.

The Army Corps knew it had made its own mistake so it contacted its Coastal Engineering Center to find out where it had gone wrong.

Chapter 30
The Paper
1973

Stabilization Changes to Sediment Flow-by Dennis Hubbard

"The physics of sand is not rocket science. It is much more complicated than that."
~Albert Einstein's advice to his son,
warning him to avoid coastal geology.

In 1973, Miles Hayes approached Dennis Hubbard with an intriguing problem, how does the Merrimack River Inlet work, and why had the Army Corps of Engineers botched things up so badly?

The Corps had a well-funded research program in the 1970's and was willing to learn from its mistakes. It often sent research problems to Dr. Hayes' lab and he passed them on to his most promising graduate students, first at the University of Massachusetts then later at the University of South Carolina.

And Dennis was one of those exceptional students. He both wrote well and loved tackling difficult problems — and the Merrimack Inlet was a doosy.

All the evidence indicated that sand should flow south through the Merrimack embayment, but on the Plum Island side of the inlet it flowed stubbornly north.

Dennis figured that a river of sand had to somehow skirt around the mouth of the Merrimack the way it did in other inlets. So he opened his paper by looking the Murrells Inlet in South Carolina. It had so little tidal flow that sand simply choked up the mouth of the inlet, impeding navigation.

That certainly wasn't the case with the Merrimack River. It came cascading out of the White Mountains of New Hampshire and rushed into the Atlantic that often had its own 12-foot tides and 20-foot waves. All that water flowing out of the river pushed sand far out into what geologists called the flood tide delta and locals called the harbor bar.

Another system Dennis described was Kiawah Inlet, also in South Carolina. In that system, the longshore currents were so strong that they had been able to push sand across the mouth of the river forcing it to migrate almost half a mile south before the system was able to form a harbor bar so sand could skirt around the new inlet. The whole process had taken less than 10 months.

But both of those inlets were natural systems. Dennis pondered what would happen when you added an artificial structure like a jetty to such a system. The perfect example of that was down on the nearby Cape Cod Canal. He took aerial photographs of the system that showed that the canal currents were so strong, and the jetty was so effective at blocking the flow of sand, that the beach on the north side of the canal was several hundred meters wider than the beach on the south side of the canal in Sandwich.

Chapter 30
THE PAPER

Having set the scene, Dennis turned his attention back to Plum Island. If sand was flowing south he suggested that you would expect that the beach on the northern side of the inlet would be wider than the beach on the on the southern side, like at the opening to the Cape Cod Canal.

But exactly the opposite was true. The beach on the southern side of the inlet was a good 300 meters wider than the beach on the northern side. In other words, the beaches were offset. The Plum Island side of the inlet was much wider than the beach on the Salisbury side.

Before the jetties had been built, the different widths of the two beaches had flipped back and forth as sand flowed unimpeded back and forth across the mouth of the river after major storms.

So Dennis hypothesized that 90% of the time, during calm weather, southeasterly winds created longshore currents that moved sand north along the northern end of Plum Island. But during Northeasterly storms, sand was torn off Salisbury Beach and then bypassed the mouth of the Merrimack River, building an offshore bar as far south as Southern Boulevard — half a mile away.

But when the fair weather returned, long period Southeast waves pushed segments of the offshore bar back onshore where they reattached to the beach at oblique angles in what are called transverse sand bars. These would fold into the beach and the sand would continue its voyage north until it came to the south jetty which would direct the longshore currents back out to sea where the sand would be lost to the system.

Some geologists would call this system of northerly flowing longshore currents and southerly flowing offshore currents a sand cell, closed when it is in a natural state but leaky when jetties shoot sand offshore and out of the system.

Finally Dennis used an elaborate mathematical formula to show that a storm packing eight foot high waves coming in from the Northeast would start ripping large amounts of sand off of Salisbury Beach and around the harbor bar to Plum Island.

The paper was a brilliant piece of work that implicitly showed that the Corps' repairs had made it impossible for the sand to get through the jetty to rebuild North Point, putting close to 80 homes in danger.

It would have also explained why the jetty was eventually going to be undermined and fall into the river but the Blizzard of 1978 did that first. In doing so the storm "disheveled" the boulders enough so that sand could flow through the jetty again, and North Point started to regrow as it had before the repairs.

So, it had taken a blizzard to cover up the Corps mistake. But what happened to the paper that had figured it all out?

As soon as the problem disappeared the paper slipped back down into the morass of scientific literature and became buried as effectively as the woolly mammoth's remains had been buried in the Knobbs Beach drumlin. But like the Matriarch's bones, would Dennis' paper ever see the light of day again?

Chapter 31
Sewer Lines on the Seashore
2004

Pumping sewerage out of Plum Island Homes, January 2015

By the early 2000's almost everyone on Plum Island was suffering from bouts of diarrhea and the smell of sewage hung heavily in the night air when the wind didn't blow. House lots had become so small due to overbuilding and moving houses back from the shore, that septic tanks were leaching into water wells. Essentially, people were drinking their neighbors' sewage.

This would probably have been a good time to let the real estate market gradually reduce the population of the island, or for zoning ordinances to increase the size of house lots. In the short term such measures would have reduced tax revenues and real estate profits, but in the long term it would have increased town income and land values. But bigger forces were also in play.

The EPA had passed the Clean Water Act in 1972, followed in 2002, by the Beach Environmental Assessment and Coastal Health Act, or BEACH. The acts required that Massachusetts force Newbury and Newburyport to enter into consent decrees with the new regulations.

Of course the island communities couldn't comply because their lots were too small. So town, state and federal officials met behind closed doors to come up with a plan. It was another doozy. Bury water and sewer lines beneath the barrier beach.

The reason you don't want to bury water and sewer lines under a barrier beach is that storms tend to break them in two. Plus, barrier beaches are underlain with salt water that rises and falls with the tides. This pushes a shallow lens of fresh water to the surface where it often floods roads, streets and low-lying areas. It can also contaminate people's wells because they get their drinking water from that same shallow aquifer.

The same sort of thing could happen with the water and sewer lines. Corrosive salt water would surround them at least a few days every month. This was the reason that the only real barrier beaches that have water and sewer lines in Massachusetts were Plum Island and Salisbury Beach.

Almost every town, state and federal environmental regulation had to be bent, broken, and folded to allow the pipes to be buried, but they were, and the $30 Million dollar project went forward. Newburyport compounded the problem by installing a finicky new system that used air vents and valves to create an underground vacuum so the sewerage could be sucked from houses on Plum Island to the waste treatment plant in downtown Newburyport, almost 6 miles away. Plus, the waste treatment plant itself was built on the banks of the flood prone Merrimack River.

Chapter 31
Sewer Lines on the Seashore

Sewer systems normally use gravity, assisted by pumps, to push water and sewerage through its pipes. But apparently a sharp salesman had convinced Newburyport's mayor that Plum Island would be better served by installing his company's new Air-Vac system. He argued that with the new system you wouldn't have a massive spill of sewerage if a storm broke through one of the pipes.

It didn't really matter that Air-Vac was a complicated system that had never been tried in the North. But the salesman was persuasive; he convinced both Newburyport and Provincetown to buy the system. Of course, P'town will try almost anything.

Problems started cropping up in 2009 then peaked in 2015 when an unlucky city employee forgot to reopen a valve after completing a 2 AM routine maintenance drill. Plum Island was still reeling from 10 feet of snow that had fallen in 30 days. This made it impossible to relocate the valve before the problems started to cascade through the system. It also didn't help that the employee couldn't remember which valve he had left closed.

The candy cane shaped vents that jutted out of the ground beside everyone's homes were clogged with snow. They were supposed to let air into people's pipes so the sewage could flow out smoothly. It was like holding a can of soda upside down. The water would only gurgle out slowly, but if you pierced the bottom of the can the water would gush out with abandon.

But the vents couldn't work buried under snow. Plus many of the wells that housed the delicate valves were covered in solid ice because they had been built in low-lying areas of the island.

But the biggest problem was the environment itself. The pipes were not fastened together to guarantee a watertight seal. Instead they had been simply lain down in a shallow trough so that the end of one pipe would rest against the gasket of the next pipe. If storms or high tides caused the sand to settle, sand, salt water and grit could all infiltrate the slumped pipes and break the vacuum.

All these problems cascadedon top of each other until over 600 basements had filled with raw sewerage and islanders couldn't use their own toilets, for fear of backing up sewerage in their neighbor's homes. Police details finally had to assemble hundreds of people in convoys of 20 cars each and slowly evacuate them off the island in blizzard conditions. Many of them opted to stay in hotels where Newburyport taxpayers picked up the tab to the tune of $70,000. Had burying the water and sewer lines been Plum Island's third big mistake?

Not a good environment for pipes

Chapter 32
Geri's Loss
November 25, 2008

Geri's Loss

Geri Buzzotta put away the warm pies she had been baking for Thanksgiving morning, said goodnight to the picture of her deceased husband Mario, and fell into a fitful sleep. An hour later her grandson heard a crack in the floor below his bed and rushed to his grandmother's side.

"Grandma what was that noise?"

"Oh, you probably just heard an especially big wave. This house has weathered many a storm, now just go back to bed. Tomorrow is going to be a big day."

"No Grandma, I heard it right under my feet. I think we need to get out of here!"

Another crack and Geri was convinced, she left her house of 46 years with only her grandson, her pet Chihuahua Oliver, and the nightgown she was wearing.

When she returned the next morning Geri was blocked from returning to her house. Her lot was cordoned off with yellow tape, and a cluster of town officials mingled beside her front door. The building inspector, Sam Joslin, broke away from the group to tell her she couldn't go back into her own home.

"But Sam, I need to some clothes and the cookies I made for Thanksgiving! My only picture of Mario is in there too!"

"I'm sorry Geri, I cant let anyone back inside. The central support beam under your house has broken.

Several hours later Joslin gave the word and an excavator nudged Geri's house gently over the dune's edge and her former home tumbled into the frigid waters of the Atlantic Ocean.

"I'm sorry Geri," said Sam turning to hug his neighbor.

"Sam, I thank you honey. It wasn't your fault. I just wonder if Mario is looking down."

Mario Buzzotta had bought the old shack for Geri 43 years before. Then, with their own hands, he and Gerri had lovingly remodeled it into their ocean front dream house. But, now Geri's savings, her belongings and Mario were all gone. The house had been all she had really wanted and now it too was washing back and forth in the incoming waves.

Gerri knew the island was eroding. They had weathered storms in the '70's, 80's and 90's but this was different, the store of sand in the center of the island had largely washed away leaving houses precariously perched on the edge of the duneline. Gerri's tragedy was only the island's most recent loss to sea level rise.

Chapter 32
Geri's Loss

With such a history, officials might have used the loss of Gerri's house and the abandonment of several others as a spur to discourage development in the dunes. But Newbury and Newburyport had just spent the $30 Million dollars to bury the water and sewer lines. If the island breached as was expected in the next big storm, up to 750 winterized homes could be cut off from water, sewerage and rescue vehicles.

Instead, Geri's loss only spurred local officials on to complete a story-high seawall of sand-filled jute sandbags. The idea behind the $2.5 Million dollar project was to provide time so the Army Corps of Engineers could come in and devise a more permanent solution.

But was there a permanent solution? Plum Island is made possible by the erosion of the large dunes and drumlins in the center of the island. And that reserve was running out.

Dennis Hubbard had also found that sand was swept around the harbor bar during Northeasters. So after a severe season of erosion there would be two pulses of sand on the beach, one that had come north from the center of the island and one that had come south from Salisbury Beach.

But the southern groinfield interrupted both these systems creating what locals called a hotspot of erosion, threatening about 40 houses on Annapolis Way, Southern Boulevard and Fordham Way. This man-made hotspot was the weak point where the Atlantic would aim its knockdown blow.

Chapter 33
The Storm
March 10, 2013

The March Storm, 2013

On March 10, 2013, Kathy Connors woke from a fitful sleep. The house was shuddering with the impact of every twenty-foot wave. She usually felt safe when Bob was at home. He seemed to know what to do in almost any situation. He had designed their house to be able to survive a Category 3 Hurricane. It rested on hundred foot pilings sunk fifty feet into the beach. The only problem was Mother Nature. She didn't care about pilings, she had designed the beach to wash away during Northeasters.

The day before, the Connors watched as their neighbor's house had swayed, shivered and toppled into the surf. Last night the house on their other side had groaned, then crumpled onto the beach. It had been like listening to a large animal gasp, collapse and die. Now a mass of water-soaked rugs, chairs and mattresses surged back and forth in the incoming tide and their neighbor's refrigerator was drifting toward Portugal.

Bob had led a constant stream of cops, politicians, and camera crews through the house every day for the past few months. Kathy had to watch as their muddy boots ground sand, snow and ice into her newly polished floors and just cleaned carpets.

Now their house was the nerve center for the immediate crisis. Every morning Senator Tarr would hold a press conference on the Connor's deck. Then the cameras would turn toward their neighbors' homes and the deathwatch would begin.

Bob was always at the senator's side, shaking hands, offering food, puffing on his cigar and leaning over the deck to give instructions to the excavator operator placing 5,000-pound concrete blocks beneath his home. The state considered such devices illegal because they increased erosion on downstream lots. But Senator Tarr had urged Bob to go ahead anyway. He would cover his back.

But Bob was the man in control; in control of the message, in control of the political process, in control of the rules and regulations devised by humans to deal with the dicey proposition of living beside the Atlantic Ocean.

He had already convinced Congressman Tierney to have Congress pay the Army Corps of Engineers $12 Million dollars of the taxpayer's money to fix the Merrimack River jetty. His main argument was that it would somehow protect homeowners from erosion almost a mile away. He had also convinced his neighbors to spend thousands of dollars to scrape the beach, creating an artificial dune and build a quarter mile long seawall of sandbags.

Chapter 33
The Storm

It had only taken a single high tide to wash away the artificial dune leaving his staircase dangling 40 feet in the air, and this storm was ripping through the sandbag seawall.

The 5,000-pound concrete blocks beneath his home, wouldn't work either, because the one thing that Bob was not in control of was Mother Nature. And Mother Nature was now calling the shots in the guise of the fourth major blizzard to batter this coast since Hurricane Sandy.

Twenty-foot waves, a nine-foot high tide and a two-foot storm surge were battering the undersides of Bob and Kathy's home. The waves had already torn a gaping hole through their garage floor. The press couldn't help looking into the abyss of churning white water before climbing the stairs to Bob's deck that was shaking from the impact of every wave. It was like staring into a watery version of hell.

Bob paced back and forth across his deck as the sun was setting. He looked like the captain of a ship, the captain of his own destiny. But the captain that Bob most resembled was Captain Ahab and it was Moby Dick circling below, preparing to bash in the bottom of Bob's Plum Island home.

In the days following the storm, it became clear what had actually happened. Six houses had been lost, 7 condemned, 1 moved and 24 declared in imminent danger. Plum Island had the most concentrated area of coastal erosion of anywhere on the East Coast during the storm.

Bob' Annapolis Way neighborhood had lost thirty percent of its housing stock, and the losses would reverberate through her families for generations to come. The homeowners felt betrayed and with good reason, they had been sold a bill of goods.

In one night, nature had done to Plum Island what humans should have been doing for several years, reducing the number of houses on this fragile barrier beach. They had had their opportunity during the Seventies.

In the wake of the Ash Wednesday storm that occurred in 1962, scientists had discovered that barrier beaches are like dynamic living entities that have to be able to regrow after every storm. Town, state and federal agencies had used this information to devise environmental regulations designed to protect homeowners from erosion, and barrier beaches from overdevelopment. But the regulations contained loopholes so wide that any developer with enough money and political clout could easily circumvent the law. Bob would soon find an ally in attacking such regulations.

CHAPTER 34
The Pacific Legal Foundation
2013

Building the Seawall

Things started to get very confusing after the March storm. Meetings were held behind closed doors, agreements were made then overturned. Trucks dumped multi-ton boulders on the beach and excavators fashioned them into an irregular seawall two thirds of mile long. Then bulldozers covered that seawall with sand scraped off the public beach. But the no one saw what happened because the Newbury police had closed the beach and posted guards armed with pistols.

Normally in a situation like this, homeowners would have had to hire an expensive engineer to draw up suitable plans, present them to the town's conservation commission, and have them approved by the state, plus have a lawyer confirm that the plans didn't run counter to any federal regulations before proceeding.

If they simply went out and started building a seawall illegally, in most other communities the town would have obtained a cease and desist order, the police would have stopped any trucks carrying boulders across the town lines and the contractors would have been fined at least $250 every day their equipment was on the beach. They could have also lost their license, plus, the feds might have fined them for having heavy equipment on a beach when endangered birds like plovers and least terns were nesting.

But apparently Newbury had given the homeowners permission and the state DEP had turned them down before getting a call from the governor's office and Bob's lawyer at the Pacific Legal Foundation in California.

The Pacific Legal Foundation is one of the country's most powerful anti-environmental organizations. It was started by then governor Ronald Reagan and funded by large donations from the Koch brothers, Exxon Mobil and Phillips Tobacco. It cut its teeth supporting nuclear power plants, arguing that cigarette smoke didn't cause cancer, litigating for the right to drill oil in public lands and fighting against Title IX funding for colleges with female athletic programs.

It had deep pockets and Massachusetts couldn't afford to be sued. It was rumored that some of the homeowners' representatives had gone to the governor and he had said something that made them believe he had given permission to proceed. But they declined to give their names to the local press.

The DEP stuck to its guns insisting that the state had not given homeowners permission to build the seawalls, pointing out that they would also be illegal in almost every other coastal state. But they also said the state would not stop people from protecting their homes during the crisis, but after the erosion season passed the homeowners might be required to remove the rocks at their own expense.

Chapter 34
The Pacific Legal Foundation

But the owners decided to go ahead anyway. During construction however, they switched to putting projectile-sized rocks on some portions of the seawall, sometimes neglected to use anti-erosion fabric behind the seawall and left gaps in front of the lots where the houses had been moved or washed away. Then they used bulldozers to scrape sand off the public beach and push it onto the seawall so you couldn't see what they had done.

During the summer several people slipped while clambering over the sharp boulders because waves bouncing off the seawall had washed away the beach in front of the seawall. It was also dangerous to swim because many of the smaller rocks had tumbled down the face of the seawall and ended up below the low water mark.

As the winter progressed, the seawalls focused waves' energy into the gaps and around the ends of the wall threatening to undermine the water and sewer lines. Newbury's board of selectmen voted unanimously to declare a state of emergency and gave the homeowners permission to build the seawall again, this time filling in the gaps and extending the wall to protect the Bennett Hill cottage that had been built in the 1800s. Apparently its owner had threatened to sue because his neighbor's seawall had caused 60 feet of the sand dune in front of the cottage to wash away, as the state had predicted it would.

In the end, homeowners who had already paid about $40,000 each to build the original seawall had to pay tens of thousands of dollars more to build it again, and it looked like this would become an annual event. The lawyer from the Pacific Legal Foundation argued that the owners had the right to defend their home.

Few argued the point, but didn't he remember the time-honored principle that his clients' rights to swing their fists ended at their neighbors' noses? Sure his clients had the right to defend their homes but did they have a right to destroy their neighbor's homes and the public beach as well?

Besides, was it really worth it for homeowners to spend more than the value of their homes to build and maintain a seawall? They had already devalued the worth of their oceanfront properties because the seawall made it impossible to climb down to their very own beach. Some owners threatened to retaliate by making the beach private.

That might even win in a court of law because Massachusetts law gave private property holders ownership down to the low tide mark. It was a hold over from colonial law. But it would surely lose in the court of public opinion.

*Road where trucks had repaired jetty.
All the land to the left of this wall had eroded away in 4 short months.*

Chapter 35
Academic Politics
December 19, 2014

Trucks repairing seawall, June 2014

"Academic politics are so vicious, because so little is at stake."
-Henry Kissinger, August 8, 1977

December 19th was a bitterly cold day with no wind and low dark clouds. The remains of a sandy trail hung six feet above the beach, surrounded by huge tangles of snow fencing. Only a few months ago, huge flatbed trucks had lumbered along this road covered with steel plates. Each truck had dumped half a dozen 5-ton boulders on the beach where brightly colored yellow excavators had picked them up and deftly placed on the jetty, last repaired in 1970.

Plum Island homeowners were thrilled. They were sure that this $12—Million dollar outlay of public funds was going to solve their erosion problems, almost a mile away.

But it was clear that something had gone terribly wrong. Waves were scouring sand off the end of North Point and the dunes were eroding back thirty to fifty feet every month. During storms they could break through the dune line and run as far as 150 feet back into the dunes. They were stopping only 300 feet from 80 homes on Northern Reserve Terrace, most of them worth over a million dollars. At this rate, the ocean would be lapping at the houses' foundations in 2016 and they would be in imminent danger by 2017.

The homeowners were victims of a common misperception. Local knowledge had it that a powerful current flowed south out of the mouth of the Merrimack River carrying sediment that had built up a sandbar that focused erosion on Annapolis Way a mile away.

If this had been the case, the Merrimack would have had to been as strong as the Mississippi River. Plus, the sandbar was also thought to grow in a cyclical pattern that focused erosion on different parts of the beach in a repeatable manner. Unfortunately this was also a convenient myth.

But how had these myths come about? If you look at charts from the 1800s you can see arrows showing that currents flow from the center of the island north and from the center of the island south. Geologists call these longshore currents because they flowed parallel to the shore constantly moving sand up and down the beach. 19th century Newburyport fishermen knew these currents well because they had to row against them to reach their fishing grounds.

But back in the 1960's erosion had also become a problem, sometimes washing away as many as half a dozen homes in a single storm. Locals had brought in The Army Corps of Engineers who had hired Duncan Fitzgerald to investigate. He gave the problem to a graduate student who had used an old fashioned wind rose to show that winds blowing out of the Northeast set up currents that flowed south along this part of Plum Island.

Chapter 35
Academic Politics

This was before oceanographers had placed an extensive array of offshore wave buoys along this coast. Now meteorologists know that in the beginning of a Northeaster, short period waves do approach from the Northeast, but as the storm develops it creates larger, longer period waves that approach the beach directly from the east. There might be a few shorter period waves skidding along on top of the long period waves, but the major swell will hit the shore directly from the east.

This happens because waves grow deeper as well as higher the longer they travel over the ocean. So when northeast waves near the coast their northern end hits the ocean bottom first and friction causes them to swing around so they end up attacking the shore straight on from the east.

But based on the false assumption that the longshore currents flowed south instead of north, the Corps had gone ahead and repaired the jetty. And as we have seen, by 1973 they realized they had made a mistake so they had brought in Dennis Hubbard, who discovered that 90% of the time the longshore currents flowed north. And it was only when waves were over 8 feet high during major Northeast storms that sand from Salisbury Beach could skirt around the mouth of the Merrimack River sometimes getting as far south as Annapolis Way.

This meant that that during 90% of the time the repaired jetty was acting like a giant dam holding sand back from building North Point. The problem was, Duncan Fitzgerald's analysis was simple and Dennis Hubbard's was complicated. After all, most rivers ran north to south and water runs downhill, right?

At any rate, Newburyport's mayor, Byron Matthews, glommed onto the simpler concept and all the other local town and city officials followed like ducklings in a row. It was just plain easier to remember that all the currents flowed south rather than that those on the north flowed north and those on the south flowed south.

But academic politics also played a part. Dennis Hubbard was a lowly graduate student at publically funded U-Mass when he wrote his paper while Duncan Fitzgerald had been at privately funded, Boston University. Then, after he finished his thesis Dennis had moved on to get his PhD at the University of South Carolina and to become head of the geology department at Oberlin College.

However, Duncan Fitzgerald had stayed on at BU where he became a well-known professor famous for telling his "joke of the day," at the start of each class on beaches and dunes. He also became known as the reigning expert on all things Plum Island.

But, Fitzgerald also continued to do most of his Plum Island research for the Parker River Wildlife Refuge. They owned the southern half of the island where the currents flowed unequivocally south, so there was no incentive for him to redo his research on the north end of the island. And, because Plum Island was considered to be Fitzgerald's bailiwick, nobody else would redo the research even if money was available, which it wasn't.

So there it was, local misperception backed by out-of-date research had all fed into the convenient myth that if you just had enough money to build groins, jetties and seawalls you could beat erosion and stop the Atlantic Ocean dead in its tracks.

CHAPTER 36
The President's Day Blizzard
January 26, 2015

Track where trucks repaired jetty

On January 26, 2015, Massachusetts Governor Charlie Baker announced that starting at midnight, all motorists would have to be off the roads due to an impending blizzard. It was not a great surprise. Weathermen had been predicting all day that this would be an historic, if not biblical storm, that could drop as much as three feet of snow on both Boston and New York.

As soon as he heard the news, Bob Connors swung into action. He called his favorite Boston television station, offering to let them embed a reporter in his Plum Island home so she could be there first thing in the morning to say that the island's new seawall was working.

Bob had a lot riding on that message. He had helped convince taxpayers to pay $12 million to repair Merrimack River's South Jetty and his neighbors to spend $40,000 each, to build the $1.6 million dollar seawall — all to protect their homes.

But the interview didn't get off to a very good start. The anchors introduced the young reporter by asking how her night had been.

"Well I didn't get very much sleep. The house kept shaking every time another wave hit. These waves are over twenty feet tall and they just keep pounding and pounding on this shore. "

Bob bit his lip. Instead of saying the seawall was working, she kept talking about the size of the damn waves and that high tide was still half an hour away.

Then the reporter directed her cameraman to film on the north side of Bob's home where waves were breaking over his seawall and pouring into Annapolis Way. This was where the island's sewer lines were buried. The high tide could also raise the ground water level, inundating the underground pipes.

Bob redirected her attention back to the south side of his house, where you could barely discern a few structures in the far distance. He insisted that the houses were doing just fine and the seawall was working as planned. But the reporter didn't seem convinced and asked him how long he felt the seawall could last.

This was not going at all according to plan. Bob said he had to go inside, brusquely brushed past the videographer and disappeared into his home. The reporter covered for him by saying, "Bob has been out all morning and is getting cold."

But the weather wasn't really the problem. The problem was, Bob's message wasn't true. The seawall hadn't worked at all. In fact it had failed after every storm since it had been constructed over the state's objections in 2013.

Chapter 36
The President's Day Blizzard

The only thing the illegal seawall had done was camouflage the ongoing erosion. While waves were only overtopping the wall in some places they were actually being accelerated as they flowed between the boulders along the entire length of the wall. This turned the dune behind the wall into a slurry of sand that was carried back out into the ocean with each successive wave, and left unseen cavities behind the wall.

If the reporter came back the day after the storm, which they never do, she would have seen that fissures had formed in the sand dunes above the seawall. A few days later, she would have seen sand sliding down into the cavities, undermining the foundations of several of Bob's elderly neighbors' homes.

But that didn't matter. Viewers had been given the impression that the seawall had worked and that message was all that really mattered.

CHAPTER 37
You Cant Sue the Atlantic Ocean
March 5, 2015

The illegal seawall had collapsed endangering homes

On March 5, 2015 the Massachusetts Erosion Commission met in Ipswich to take public comments on the commission's recommendations for what people could do to protect their coastal homes. It was a colorful meeting and a sellout crowd. Bob Connors showed up with a lawyer from the Pacific Legal Foundation.

Since the foundation had cut its teeth arguing that it was OK to site nuclear power plants on some of California's most active seismic areas, it was just a hop, skip and a jump to argue that homeowners should be able to build houses overlooking the equally active Atlantic Ocean.

The foundation had already litigated against numerous coastal states arguing that a developer should be compensated for the loss of his investment if an environmental agency ruled that he couldn't build a house on an eroding beach. The amazing thing is that they won in one of the worst decisions ever made by the Supreme Court. The court decided that such an environmental ruling should be considered a "taking" under the Fifth Amendment. This gave new meaning to the expression "pleading the Fifth."

But how often do homeowners fly in an expensive West Coast lawyer to hold a gun to their state officials' heads? The lawyer graciously offered to sit down with the committee members and go over their report line by line "so you can avoid litigation." How subtle can you get?

Of course the assumption underlying the threatening argument was that the seawall his clients had built was working without a hitch. But, then a group of ragtag environmentalists requested to show some footage of the seawall taken from a drone the week before. In the spirit of full disclosure the author will admit he was one of those ragtag environmentalists.

The footage showed that the seawall had collapsed in several places, exposing foundations and leaving four houses dangling precariously over the edge of the dunes. Then it went on to show that in only 6 months time the repair of the Merrimack River's South Jetty had eroded two football fields worth of dunes off the north end of Plum Island, putting up to 80 more homes at risk.

Bob was apoplectic. He had spend the last year carefully grooming the media to present his message that the $40,000 that each homeowner had spent to build their section of the seawall and the $12 million dollars that taxpayers had spent to repair the jetty had stopped the Atlantic in its tracks.

The chairman of the committee cut our film short, but its point had been made. Erosion was still running rampant on Plum Island. But the problem was, the lawyer from the Pacific Legal Foundation was right. According to the Supreme Court any common sense regulation to protect homeowners from losing their shirts, as much as to protect the environment, was essentially null and void.

Chapter 37
You Cant Sue the Atlantic Ocean

All a homeowner had to do was hire a fancy lawyer, plead the Fifth and pay for yet another harebrained scheme to fight the Atlantic Ocean. State regulators couldn't even suggest a setback to require a homeowner to build his house on the back of his lot instead of directly on the edge of the raging Atlantic. That was supposed to be a right, guaranteed by the Constitution and supported by the Supreme Court of the United States.

The problem was, the Atlantic Ocean didn't know anything about constitutional law and couldn't give a whit about rulings by the Supreme Court. It was just going to do what it was just going to do.

I was amazed to see an excavator repairing the seawall and scraping sand off the public beach

CHAPTER 38
The Seawall Returns
April 6, 2015

It was difficult to walk down the beach at high tide

On April 6, 2015 I drove up to Plum Island to see how it had fared during the winter. Much to my surprise I discovered two large excavators and a bulldozer on the beach. The shore reverberated with heavy thuds as trucks dumped several ton boulders on the sand. The excavators would then swing into action picking up the giant boulders and placing them on the half mile long seawall. After that, the bulldozer covered the boulders with a thick layer of sand that it had scrapped off the public beach.

It was especially surprising since only a month before Bob's lawyer had been arguing how well the seawall had done its job. But now, sharp, slippery boulders were already collapsing off the seawall and people had to clamber over them at high tide because the seawall had also steepened the profile of the beach.

I decided to find out when the Newbury Conservation Commission had given permission to do the work. The last time this had been done was under a state of emergency after the wall had failed in 2013, and again just before the arrival of Hurricane Sandy.

I hadn't heard that the board of selectmen had voted to declare a state of emergency, but perhaps I had not been paying any attention. But nobody had posted the minutes of the conservation commission meetings for the past few months.

Then I e-mailed the Newbury Conservation agent but nobody answered my message so I called the state's Department of Environmental Protection. They also had no record of any permissions being given to do the work.

This seemed particularly odd. Apparently no town or state official knew about this work that would continue five days a week for the rest of the month. The work had to be in violation of the town and state's own environmental protection regulations and probably the Federal Endangered Species Act that prohibits heavy equipment from being on the beach when piping plovers are nesting.

You could also see that the repair wasn't working. On succeeding days I watched the sand wash away revealing the large new boulders underneath. If people already had to clamber over the seawall in April, what would it be like by the summer?

The foundation of one of the houses had also cracked and slumped down more than a foot, and the ocean had torn staircases off of several people's homes and strewn nail studded pieces of lumber on the beach and in the water.

Unfortunately the work was also being done in piecemeal fashion so it was leaving big gaps in the wall where houses were already missing or where their owners didn't want to flaunt the law. They would be penalized in future storms because waves would wrap around the repaired seawall and slam into their lots with increased intensity, in a process that engineers call "end scour."

Chapter 38
The Seawall Returns

Of course it also meant that people couldn't get to their own beach without scrambling down the seawall festooned with their own signs saying "private property" and "keep off the rocks." Plus they couldn't swim because the public beach was littered with slippery sharp boulders that had collapsed off their private seawalls.

It seemed like a strange way to protect your beachfront home. Why not just raise it up on pilings or move it across the street as homeowners had done before?

Chapter 39
North Jetty
September 5, 2015

During the summer of 2015 lifeguards at Salisbury Beach State Park had to rescue 105 swimmers from drowning. Usually they would only have to save 25 people. So, why did they have to make 4 times more rescues than normal than on any other beach in Massachusetts?

The answer lay right beside them; where cranes, barges, excavators and trucks were busily placing 5-ton boulders on the Merrimack River North Jetty. Before this repair, energy from Northeasterly waves would be dissipated as the waves broke through low spots on the jetty. But now the jetty presented a straight, impenetrable wall without holes so the waves would ricochet off, scour out great scallop shaped portions of the beach and then return back out to the ocean as dangerous rip tides.

Even on calm days the area that has been scalloped out by storms could harbor these currents that develop quickly and shoot swimmers directly out to sea. It is impossible to swim against a rip tide and veterans know you have to turn and swim parallel to the beach in order to find water calm enough to return back to shore. But even veterans can drown in the grip of a rip tide.

But how did this situation come about? The same way it came about when they repaired the South Jetty on the other side of the river. If the Army Corps of Engineers had done a simple literature search, they would have discovered Dennis Hubbard's paper that explained why similar problems occurred after the jetties were repaired in 1970.

In the paper he explained that when you have a Northeasterly waves they ricochet off the jetty leaving the giant scalloped gouges out of the beach that direct the longshore currents offshore as so-called rip tides.

The corps should have made their own regional sediment flow analysis or at least done a literature search before undertaking the two $12 Million dollar jetty repairs. If they had, they could have avoided both the rapid erosion on North Point and the creation of dangerous rip tides at Salisbury Beach.

Chapter 40
The Matriarch's Message
August 31, 2013

The Matriarch's Message

Back in 2013 the Matriarch had reemerged on the last day of August. But she was not found jutting out of the sand on Plum Island but in the pages of a Long Island newspaper. Apparently, Homeland Security and the former Army Chemical Corps had been studying animal diseases on a pork chop shaped piece of land off New York they also called Plum Island.

But in 2008, Congress had voted to relocate this bioterrorism laboratory to of all places, Manhattan, Kansas. Then they planned to sell the former site to be developed. But who would want to live on an island riddled with Anthrax and Rinderpest spores? First the government had to conduct an exhaustive study to see whether the research had left behind any contaminants.

During the course of the study, an environmental group had found an old newspaper article about the discovery of the remains of a woolly mammoth on Plum Island in 1879. They were elated because it meant that more archeological studies would have to be done before the island could be sold.

The New York media ran with the story, but it never quite rang true. The article said the massive skull; backbone and leg had been discovered near a life saving station. But there had never been a life saving station on Plum Island in New York!

Eventually a librarian discovered that the article had originally appeared in the old Newburyport Herald and was about Plum Island Massachusetts, not Plum Island, New York. A local Long Island paper had simply lifted the article word for word, as was their wont in those days.

Once the story shifted back to New England more questions emerged. Exactly where had the bones been found? Where were they now, and what significance did they have for modern day Plum Island?

The article said that "three gentlemen" had found the skeleton protruding from, "an elevation of sand know as 'Brothers' Beach,' 150 feet long and 50 feet high, one of the largest sand hills on the island. Latterly the winds have blown it away, so that the sand dune has lowered to a height of only a few feet."

The problem was nobody knew where Brothers Beach had once been. Jerry Klima, a former selectman from Salisbury who had studied numerous historic maps of the island thought that it was probably one of the places where people congregated after the Civil war to enjoy their weekend picnics.

But the original article had also said that the site was near a life saving station. In 1879 there was only one life saving station near the center of the island and that was at High Sandy called Knobbs Beach in the 1800s. But by 2013 the original site was probably a hundred feet offshore and under 10 feet of water.

Chapter 40
THE MATRIARCH'S MESSAGE

The other mystery was what had happened to the Matriarch's venerable bones. The usual suspects, the Peabody Essex Museum in Salem, and the Peabody Museum at Andover Academy had never heard of the Matriarch. The curator of paleontology at Harvard's Museum of Natural History, Dr. Agassiz's old haunt, checked through her database and also found nothing. But she suggested that if the bones were ever unearthed they would make a stunning exhibit in the Parker River Wildlife Refuge's new auditorium near the entrance to Plum Island.

The article had also mentioned that the bones were already crumbly from being interned so long in the dune's drying sands. The curator was sure the remains of the Matriarch had probably crumbling away in the attic of one of the old sea captain houses on High Street.

But perhaps the Matriarch's most important message was how quickly the world could change. She had watched the glaciers retreat and the oceans rise almost 30 feet within her short lifetime. She had seen humans extirpate her continent's mega fauna in just a few fleeting generations. She had felt the life threatening warming as the Ice Age came to a rapid close.

Now the planet was warming as fast as it had at the end of the last Ice Age and the coasts were retreating faster than ever before. Would another woolly mammoth emerge from the dunes of Plum Island as the Matriarch had in 1879? Would she have any better luck convincing us of the folly of our ways? Only time would tell…

CHAPTER 41
The Last Folly?
September 18, 2015

In mid-September, the Army Corps of Engineers presented a plan to dredge 370,000 cubic yards of sediment out of New Hampshire's Piscataqua River and dump it in the water off Plum Island. The idea was that waves would winnow sand out of the mixture of mud and rocks and wash it ashore to protect Plum Island's endangered homes.

But would it be just the latest in a long line of Plum Island debacles. We have seen that groins built in the 1960's created a hotspot of erosion in the Annapolis Way- Southern Boulevard neighborhood. An illegal seawall built in 2012 failed in 2013, 2014 and 2015. Merrimack River's South Jetty, repaired in 2014, put 80 homes at risk on North Point. And Merrimack's North Jetty repaired the following year created deadly rip tides on Salisbury Beach.

All of these projects were done without proper sand transport studies and apparently without even reference to previous research. If a proper literature search had been conducted, the Corps would have discovered Dennis Hubbard's nugget of shining research amongst all the scientific dreck.

Dr. Hubbard pointed out that sand flows north along the beach during calm weather and south again during Northeasters, creating a giant self contained sand cell that works to keep sand in the system.

But the problem with the island's groins and jetties is that they interfere with this flow of sand that wants to repair the beach naturally. So if the Corps doesn't open these structures, any sand pumped onto the beach will simply be jettisoned back out to sea near the Merrimack River so it cannot be returned naturally to the Annapolis Way area. And the Corps will have flushed $1.5 Million dollars down the drain in another vain attempt to protect half a dozen houses on Annapolis way.

The problem is that now, those parts of the beach above the groins have about 80 feet of sand between the waterline and the dunes, but those parts below the groins have no beach at high tide. But if you open the groins and jetties the entire beach would widen by 60 to 100 feet, and sand will be able to repair North Point again. This will be better for both the houses and the beach.

In time an organization like the Merrimack Beach User's Association might hire a dredge to enhance the natural system but pumping sand from where it tends to accumulate at the northern end of the sand cell back to the center of the island. This will enable the island's longshore currents to distribute it evenly along the four-mile long stretch of the beach.

If we do this we will become like organic farmers that have learned how to work with the earth's natural ability to recycle nutrients back into the soil rather than to use harsh petroleum based herbicides and fertilizers to deplete it. These methods wont stop the Atlantic in its tracks. That is impossible. But it will put us on a path of learning that we can work with the natural systems of the ocean instead of waging a fruitless war against it.

Epilogue
A Golden Opportunity
2015

Newbury and Newburyport now have a golden opportunity to step back from trying to solve the day-to-day problems of sewers and seawalls to visualize how they would like Plum Island to look 5, 10, 20, even 50 years from now.

The National Wildlife Foundation, the Merrimack Valley Regional Planning Authority and several other environmental groups have received a $2.9 million grant to help towns develop plans to make their communities more resilient.

The biggest problem they will have to solve is money. Right now towns like Newbury get 40% of their tax revenue from only 33% of their land and houses built in the fragile dunes that are in the process of washing away due to rising seas and more frequent storm activity.

There are several way to replace these tax revenues that are being swept away year-by-year.

1— The first thing that Newbury and Newburyport can do is to provide off-island parking and transportation to the beach. Crane's Beach, just to the south of Plum Island in Ipswich receives $700,000 a year from parking tolls and Salisbury Beach just to the north of Plum Island state to state receives $250,000 a year.

Overhead to run such parking lots is low. All you have to do is hire a few high school students to collect the parking fees, or have meters that do it automatically.

And, this would help return Plum Island to the warm welcoming beach community it once was, instead of a place for year round residences sporting signs saying things like "Do not park," "Tow Away Zone," "Private Property" and "Keep off the Rocks."

2 – The towns could also charge a small toll to get on the beach. The entrance to Plum Island is already a turnpike; all you would have to do is set up a booth to collect modest tolls the way they did when the bridge was first built and still do on similar islands in places like Florida.

3— Provide shuttles from in-town parking lots and the Newburyport train station. A pilot project to this effect was started in 2015. The increased numbers of people using the beach would provide a multiplier effect as they buy food, beverages, beach equipment and fishing tackle. You could then use some of this revenue to bring back the lifeguards, sanitary facilities and seasonal food concession stands that used to dot the island.

4 — Gradually get rid of the groins and seawalls. This would allow the beach to widen by 60 to 100 feet. Now the beach is so narrow beach goers have to clamber onto the seawall's boulders in order to walk down on the beach at high tide. And swimming has become treacherous in the intertidal area below the seawall.

5 – Have the state gradually buy out threatened homes. There is already a bill in the Massachusetts Senate to provide $20 million dollars to buy out vulnerable homes. The empty lots could then be replanted and made into mini parks and scenic overlooks.

Before erosion really started picking up in 2008 there was a solid barrier of imposing homes on top of the dunes; now there are several places that have spectacular views where the houses washed away. This has increased the value of the second row of dwellings.

If we don't act proactively to do these things nature is going to do it for us in a most unpleasant manner. I think we have to face the fact that none of the hard engineering solutions have worked. We have to start working with nature or as some Plum Island residents are starting to say, "I have spent 60 wonderful years in this house, but now the ocean wants it back and it is time to leave".

Footnotes

Prologue
Rogers, Dave. "Storms take a toll." Newburyport News. 2/17 2015.
Hendrickson, Dyke. "Conditions are improving on Plum Island." Newburyport Daily News. Feb 21, 2015
Samenow, Jason. "Ash Wednesday Storm in 1962: 50 year anniversary." Capital weather Gang, Washington Post. 3/6 2012.

Chapter 1
Sargent, William. "A Year in the Notch." UPNE. Lebanon, NH 2000.

Chapter 2
Sargent, William. "A Year in the Notch." UPNE. Lebanon, NH 2000.
Twain, Mark. "Life on the Mississippi." 1883

Chapter 3
Flannery, Tim "The Eternal Frontier; An ecological history of North America." Atlantic Press. NY 2001
Sargent, William. "Beach Wars." Strawberry Hill Press 2013.

Chapter 4
Fitzgerald, Duncan et.al. New England Technical Reports 1993
Hein and Stone. "Ice Water and Wind; A source to sink view of the processes that have shaped Northern Massachusetts." 2012.

Chapter 5
Sargent, William. "Shallow Waters; A Year on Cape Cod's Pleasant Bay." Houghton Mifflin. Boston, 1980.

Chapter 6
Harris, Gordon. "Stories From Ipswich and the North Shore." Wordpress.com.

Chapter 7
Oertel and Kraft J.C. "New Jersey and Delmarva Barrier Islands."
Marta, Margaret. "Laurentide glaciation of the Massachusetts coast." ES767 Quaternary Geology Fall 2008.
Gerkin, James. "Sea Level along NE rose 4 inches in just 2 years. Huffington Post 2/25 2015

Chapter 8
Horwitz, Tony. "How Much do we really know about Pocohantas?" Smithsonian Magazine. November, 2013.
"The Pocahontas Myth." Powhatan Kenape Nation website.
Sargent, William. "John Rolfe steals Stolen Tobacco, Bermuda 1609." Unpublished MS,
"John Smith" Biographies-explorer. Biographies.com.

Chapter 9
Wood, William. "New England Prospect." 1634.

Chapter 10

"Publications of the Ipswich Historic Society." Ipswich.

Lindborg, Kristina. "Natural History of Boston's North Shore." UPNE, Lebanon, NH 1993.

Weare, Nancy. "Plum Island, The Way it Was." Newburyport Press. 1993.

Chapter 11

Weare, Nancy. "Plum Island, The Way it Was." 1993.

Chapter 12

Weare, Nancy "Plum Island, The Way it Was." 1993.

"The Great Storm of 1717." Wikipedia.

Holthaus, Eric. "Boston's Astounding Amount of Snow, a 1 in 26,315 year occurrence." Capital Weather Gang. The Washington Post, 2/25 2015.

Chapter 13

Weare, Nancy. "Plum Island, the Way it Was." 1993.

Sargent, Willliam. "The House on Ipswich Marsh." UPNE Lebanon, NH 2003

Sandy Point State Reservation Wikipedia.

Chapter 14

Weare, Nancy. "Plum Island, The Way it Was." 1993.

Behbehani. "The role of Cotton Mather in Colonial America." Microbiology.

Monagan, David. "Isolated Reminders of Old Epidemics."

NY Times 4/9 2000.

Chapter 15

Weare, Nancy. "Plum Island. The Way it Was." 1993.

Smith, E. Vale. "History of Newburyport." Brambel and Moore. Boston. 1854.

Chapter 16

Sargent, "Beach Wars," Strawberry Hill Press, 2013.

Smith E, Vale. "History of Newburyport." Boston. 1854

Weare, Nancy. "Plum Island, The Way it Was." Newburyport Press 1993.

Chapter 17

Weare, Nancy. "Plum Island, The Way it Was." 1993.

"Merrimack River." Wikipedia.

Chapter 18

Weare, Nancy. "Plum Island, the Way it Was." 1993.

Chapter 19

Pettingell, Warren. "The Pettingells. A New England Family."

Weare, Nancy. Plum Island. "The Way it Was." 1993.

Winslow Pettingell, personnel communication, who told me that, of the 109 different ways to spell Pettingell this is the correct one.

Chapter 20

Weare, Nancy. "Plum Island, The Way it Was." 1993.

Chapter 21

Macone, John. "Mammoth mysteries." Newburyport Daily News. Aug 30, 2013.

Chapter 22

Footnotes
A GOLDEN OPPORTUNITY

Waters, Thomas Franklin. "Ipswich in the Massachusetts Bay Colony." In Stories from New England. Wordpress.com.

CHAPTER 23

Sargent, William "Just Seconds from the Ocean." UPNE Lebanon, NH. 2007.

CHAPTER 24

Weare, Nancy ibid.

Sargent, William, "Beach Wars" UPNE, Lebanon, NH 2013.

CHAPTER 25

Weare, Nancy ibid,

Nelson, Liz Newburyport "Stories from the Waterside." In "Stories from Ipswich and the North Shore." Worpress.com.

CHAPTER 26

Sargent, William Beach Wars" Strawberry Hill Press. 2013.

Griscom, Ludlow. "Plum Island and its bird life."

CHAPTER 27

Carson. Rachel. The Parker River Wildlife Refuge. 1946.

CHAPTER 28

Hendrickson, Dyke. Camp Sea Haven Reunion Tomorrow." Newburyport Daily News 7/27 2013.

History of poliomyelitis, Wikipedia.

CHAPTER 30

Hubbard, Dennis K. "Changes in Inlet Offset due to Stabilization." Coastal Engineering Proceedings No 15. 1976.

Dennis Hubbard, personal communication.

CHAPTER 31

McKabe, Kathy. "A costly sewer system failure on Plum Island." Boston Globe. June 12, 2015.

Rogers, Dave. "On Plum Island, frustration as sewer failures persist." Newburyport Daily News 2/26 21015.CHAPTER 32 GERRI'S LOSS

Sargent, William. "The View from Strawberry Hill." Strawberry Hill Press. Ipswich. 2013.

CHAPTER 33

Sargent, William. "Islands in the Storm." Strawberry Hill Press. Ipswich. 2014.

CHAPTER 34

Sargent, "Islands in the storm." Strawberry Hill Press, Ipswich, 2014.

CHAPTER 35

Macone, John. "Starttling déjà vu." Newburyport Daily News 2/22 2013.

Hendrickson. Dyke. "PI's Evolving Role." Newburyport Daily News. 6/9 2014.

CHAPTER 36

Ashford, Ben. "Plum Island. Where Recent record Snowfall has left basements flooded with sewage." Daily Mail, London. 4/10 2015.

CHAPTER 39

Sargent, William. Unpublished MS.

Gaziano, Todd. Protecting your home v. letting it crash into the sea. Pacific legal Foundation. Liberty Blog 4/13 2015.

Hendrickson, Dyke "Coastal Erosion and its Impact." Newburyport Daily News. 4/10 2015

Chapter 40

Sargent, William. Unpublished MS.

Epilogue

Solis, Jennifer. "Task Forces formed to study impact of coastal storms. Newburyport Daily News. May 6, 2015.

CPSIA information can be obtained at www.ICGtesting.com
Printed in the USA
BVOW08s0506070616
450721BV00001BA/5/P